World Petroleum
Resources and Reserves

Also of Interest

Oil Strategy and Politics, 1941–1981, Walter J. Levy, edited by Melvin A. Conant

International Dimensions of the Environmental Crisis, edited by Richard N. Barrett

Critical Energy Issues in Asia and the Pacific: The Next Twenty Years, Fereidun Fesharaki, Harrison Brown, Corazon M. Siddayao, Toufiq A. Siddiqi, Kirk R. Smith, and Kim Woodard

† *Energy Futures, Human Values, and Lifestyles: A New Look at the Energy Crisis,* Richard C. Carlson, Willis W. Harman, Peter Schwartz, and Associates, SRI International

† *Living with Energy Shortfall: A Future for American Towns and Cities,* Jon Van Til

Transitions to Alternative Energy Systems, edited by Thomas Baumgartner

Alcohol Fuels: Policies, Production, and Potential, Doann Houghton-Alico

Energy Transitions: Long-Term Perspectives, edited by Lewis J. Perelman, August W. Giebelhaus, and Michael D. Yokell

OPEC: Twenty Years and Beyond, Ragaei El Mallakh

Indonesia's Oil, Sevinc Carlson

OPEC and Natural Gas Trade: Prospects and Problems, Bijan Mossavar-Rahmani and Sharmin B. Mossavar-Rahmani

Renewable Natural Resources: A Management Handbook for the Eighties, edited by Dennis L. Little, Robert E. Dils, and John Gray

† Available in hardcover and paperback.

A Westview Special Study

World Petroleum Resources and Reserves
Joseph P. Riva, Jr.

Up to 1965 the world produced and consumed only 10 percent of the oil available on this planet; between 1965 and 2040 we will use up 80 percent of the remaining reserves, leaving only 10 percent of the resource for the years to follow. Clearly, the epoch of petroleum is a transitory one. Nevertheless, petroleum is at present the most important component of the energy base supporting the industrialized world. This book describes and analyzes the geological basis for the current world petroleum situation.

Mr. Riva explains the formation and accumulation of conventional and unconventional oil and gas, the methods used by geologists in the search for petroleum and petroleum-containing basins, and techniques for petroleum production. He then discusses the uneven distribution of the world's oil, focusing on the Arabian-Iranian basin, which contains half of the world's known recoverable reserves, and examines the petroleum prospects in several distinctly different areas of the world. The United States is presented as an example of an area in general decline, already exhaustively explored. In contrast, the case study of the Soviet petroleum industry and a geological assessment of Soviet production prospects show a region at the peak of its oil production, with its decline about to begin. He chooses Indonesia as the focus for a typical Southeast Asian petroleum history and develops a profile of Mexico's petroleum situation as an example of an area with increasing production potential. Mr. Riva concludes with an assessment of the prospects for future world petroleum discoveries and a geologically based estimate of the earth's total original stock of recoverable petroleum.

Joseph P. Riva, Jr., a geologist and specialist in earth sciences, is head of the Oceans Team at the Congressional Research Service of the U.S. Library of Congress (Science Policy Research Division). Among his numerous publications are *Secondary and Tertiary Recovery of Oil* and *Energy from Geothermal Resources*.

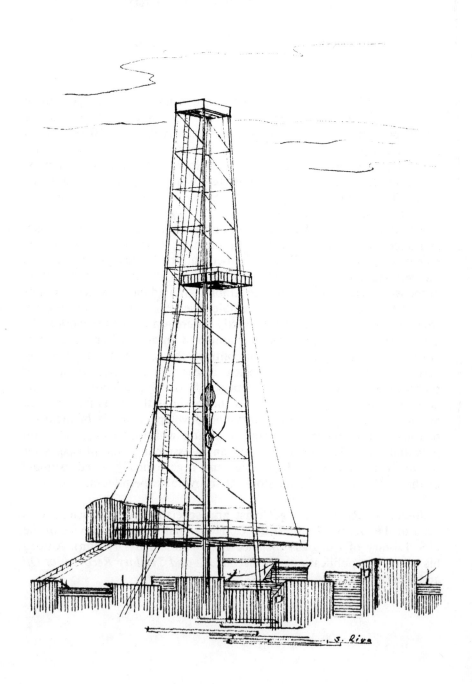

World Petroleum Resources and Reserves

Joseph P. Riva, Jr.

Westview Press / Boulder, Colorado

A Westview Special Study

Illustrations by A. Susanne Riva.

Published in 1983 in the United States of America by
  Westview Press, Inc.
  5500 Central Avenue
  Boulder, Colorado 80301
  Frederick A. Praeger, President and Publisher

Library of Congress Cataloging in Publication Data
Riva, Joseph P.
  World petroleum resources and reserves.
  (A Westview special study)
  Bibliography: p.
  Includes index.
  1. Petroleum. 2. Petroleum—Reserves. I. Title.
TN870.R53 1982   333.8'23   82-13625
ISBN 0-86531-446-2

Printed and bound in the United States of America

To Susanne

Contents

Tables

Figures

Preface

This book ends with a table that contains three numbers. The numbers represent a geologically based estimate of the earth's original stock of recoverable conventional petroleum, the product of 570 million years of natural processes. In the table, the total world ultimate recovery of conventional petroleum is estimated as 1,953 billion barrels of oil, 428 billion barrels of natural gas liquids, and 11,328 trillion cubic feet of natural gas. Estimates, of course, are always subject to modification as new information becomes available and new understanding evolves; however, we know enough about the world petroleum resource and its origins, migrations, and accumulations to put its magnitude into perspective.

The preceding numbers are large and have important implications. They indicate that the earth has been stocked with petroleum only for the short term. It took 109 years for the world's population to consume 200 billion barrels of oil; however, it took just ten years to consume the next 200 billion barrels. Twenty-two percent of total world oil stocks have already been consumed. World oil production is currently about 1 percent of the total world stock per year. If this level of production continues, the middle 80 percent of the world's ultimate oil supply will be consumed in about seventy-five years (the period between 1965 and 2040). World production is expected to increase, however, which will result in an even shorter "age of oil energy."

About 59 percent of the world's total oil has been discovered. More importantly, the remaining 41 percent will be even more expensive to find and recover. Already the capital costs per daily oil production in some areas of the North Sea are forty times the costs in the Middle East. Over 11,000 man-years are required to construct the largest of the North Sea gravity production platforms; the cost would exceed that of a nuclear power plant or oil refinery. The guyed tower constructed for oil production in 1,000 feet of water in the Gulf of Mexico will recover oil at about sixty-five times Middle East production costs. As oil production moves into the frontier basins in deeper waters or under Arctic ice, these costs will further escalate and be reflected in the world economy.

Natural gas has long been considered a nuisance by-product of oil production and often has been flared. For the past twenty years, however, rapidly increasing world gas reserves and the increasing price of oil have resulted in the recognition of gas as an extremely important energy resource. Because natural gas is more ubiquitous than oil in sedimentary basins, and past prospecting has concentrated on oil, the world's ultimate gas recovery may eventually about equal ultimate oil recovery (on a Btu basis). If this proves true, about 39 percent of the world's gas remains unaccounted for, but it probably exists in nonassessed basins, below Arctic permafrost, or below the oil window in deep basins, or it was bypassed by previous oil discoveries or flared during past oil production. The world stock of gas will last longer because the utilization of natural gas in volume lags behind the use of oil. If the use of gas approaches the use of oil on an equivalent energy basis, its length of service will be equally brief.

In any event, on a time scale within the span of prospective human history, the utilization of petroleum as a major source of energy is a transitory affair, but an affair of profound importance.

Joseph P. Riva, Jr.
Great Falls, Virginia

Unit Conversions

Length

1 inch = 2.540 centimeters
1 foot = 0.305 meters
1 mile = 1.609 kilometers

Area

1 square inch = 6.451 square centimeters
1 square foot = 0.093 square meters
1 square mile = 2.590 square kilometers

Volume

1 cubic foot = 0.028 cubic meters
1 barrel = 0.159 cubic meters
1 cubic mile = 4.166 cubic kilometers

Pressure

1 pound per square inch = 0.070 kilograms per square centimeter

Temperature

Degrees Fahrenheit minus 32 $\times$ 5/9 = degrees Celsius

Energy Units

1 barrel of oil contains energy equivalent to 5,800 cubic feet of natural gas
7.3 barrels of oil = 1 metric ton of oil (world average)

1
The Occurrence of Petroleum

Oil is generated at depths of about 2,500 to 16,000 feet, at temperatures between 150 and 300 degrees Fahrenheit, from predominantly plant debris in sedimentary rocks. This special environment is called the "oil window."

The decomposition of organic matter in an oxygen-poor environment, with the aid of anaerobic bacteria, results in the formation of methane. Organic matter is ubiquitous, and so, therefore, is natural gas. If all the natural gas could be collected, it could provide most of the world's energy for hundreds of years. Unfortunately, like gold in sea water, most of natural gas is too diffuse to be of commercial value. In the course of geologic time, almost all natural gas reaches the earth's surface and is lost.

Liquid and gaseous hydrocarbons are intimately associated in nature, and it has become customary to shorten the expression "petroleum and natural gas" to "petroleum" when referring to both. Natural gas is often dissolved in oil at the high pressures existing in a reservoir; it can also be present as a gas cap above the oil. Such natural gas is termed associated gas. There are, in addition, reservoirs that contain gas and no oil. This gas is termed nonassociated gas.

Organic matter and natural gas are ubiquitous, but commercial oil and gas fields are not. The existence of a petroleum field depends upon four essential conditions that must be fulfilled in sequence. First, the formation of a petroleum deposit requires a source rock containing sufficient organic matter that has been subjected to an appropriate time of burial and geothermal gradient to generate petroleum. Next, a permeable reservoir rock must have been connected to the source rock by a channel suitable for fluid migration by buoyancy or hydrodynamics. Third, a closed trap, like an inverted container, must have been formed in which the petroleum could accumulate. The trap is capped by rock exhibiting a general decrease in porosity that acts as a seal. Finally, the filled trap must have been preserved intact for relatively long periods of geologic time.

Source rocks are not uncommon, reservoir rocks are less common, and traps are rather rare. It is the sequential occurrence of the three and the subsequent preservation of the filled trap that are essential for the existence of a commercial deposit. Oil and gas accumulations are relatively scarce; large deposits are very scarce. This suggests that the conditions essential for the formation and accumulation of commercial petroleum deposits have only infrequently been satisfied in the required order.

The Origin of Fossil Fuels

Oil, gas, and coal originate from plant debris and are subject to the same geological processes of bacterial action, burial, compaction, and geothermal heating. There are, however, basic differences between the modes of petroleum formation and those of coal formation. Coal is found at its site of deposition as a solid, relatively pure, massive organic substance. Oil and gas are fluids that migrate readily from their place of origin through permeable rocks. The main organic precursor material of petroleum compounds is kerogen, an insoluble organic residue that is finely dispersed and intimately mixed with the mineral matrix in petroleum source beds, but usually amounts to less than 2 percent of the total rock mass. Most coals are remnants of higher, terrestrial plants; the kerogen of petroleum source beds is usually derived from lower aquatic plants and bacteria, although terrestrial organic matter can produce gas. Most petroleum source beds were deposited in marine environments and most coals were formed under nonmarine conditions.

The generation and occurrence of petroleum can be related to three main stages of thermal maturity of the organic material in sedimentary rocks.

Immature Stage (Diagenesis)

The first stage is dominated by biological activity and chemical rearrangement, which convert organic matter to kerogen, the source of most of the hydrocarbons generated in the later stages. During the first stage, biogenic methane is the only commercial hydrocarbon generated in significant volumes, but small amounts of other hydrocarbons, that have been derived from the mild degradation of organic tissue may be present. The production of biogenic methane gas is part of the process of the decomposition of organic matter by microorganisms. These methane-producing microbes are strictly anaerobic and cannot tolerate even traces of oxygen. They are also inhibited by high concentrations of dissolved sulfate, which restricts methane production. Thus, biogenic gas generation is confined to certain environments that include poorly drained swamps and bays, some lake bottoms, and marine sediments beneath the zone of active sulfate reduction. Gas of predominantly biogenic origin constitutes more than 20 percent of the world's gas reserves.[1]

Mature Stage (Catagenesis)

As a result of burial by continuing sedimentation, increasing temperature, and advancing geological age, the full range of petroleum hydrocarbons is produced from kerogen and other precursors by thermal degradation and cracking. Depending upon the amount and type of organic matter, oil generation occurs during the second stage. Oil production is often accompanied by the generation of significant amounts of thermal methane gas.

In areas of higher than normal geothermal gradient, the oil window occurs at shallower depths in younger sediments but is narrower. Maximum oil generation occurs between 6,500 and 9,500 feet, depending upon the geothermal gradient. Below 9,500 feet, primarily wet gas (gas containing liquid hydrocarbons called natural gas liquids) is formed.

About 90 percent of the organic material in sedimentary source rocks is ubiquitous kerogen. This dispersed, dark-colored, insoluble product of bacterially altered plant and animal detritus is variable in composition and consists of a range of residual materials whose basic molecular structure takes the form of stacked sheets of aromatic hydrocarbon rings in which atoms of nitrogen, sulfur, and oxygen also occur. Attached to the ends of the rings are various hydrocarbon compounds, including normal paraffin chains. The mild heating of the kerogen in the oil window of a source rock over long periods of geologic time results in the cracking of the kerogen molecules and the release of the attached paraffin chains. Further heating, perhaps assisted by the catalytic effect of clay minerals in the matrix of the source rock, may then produce soluble bitumen compounds followed by the various saturated and unsaturated hydrocarbons, asphaltenes, and others of the thousands of hydrocarbon compounds that make up normal crude oil mixtures.[2]

Post-Mature State (Incipient Metamorphism)

At the end of the mature stage, below about 16,000 feet, depending upon geothermal gradient, the kerogen becomes highly polymerized, condensed in structure, and chemically stable. In this environment, crude oil is no longer stable and the main hydrocarbon product is thermal methane gas. The thermal gas is the product of the cracking of existing liquid hydrocarbons. Those hydrocarbons with a larger chemical structure than methane are destroyed much more rapidly than they are formed. Thus, in the sedimentary basins of the world, comparatively little liquid hydrocarbon is found below about 16,000 feet, but deep basins with thick sedimentary sequences do have the potential for deep gas production.[3]

Source Rock Maturity Assessment

Modern techniques for determining the maximum temperature reached by a potential source rock during its geological history are of value for the estimation of the maturity of the contained organic material and thus the extent to which hydrocarbons are likely to have been generated. Such assessments of the viability of potential source rock can be used in estimating the oil potential of an area. Also, the maturity of the source rock organic material will indicate whether a region is gas prone, oil prone, both, or neither. An immature source rock will produce only biogenic gas, while mature source rocks will usually generate both oil and associated gas with gas liquids. Deep postmature source rocks will produce only thermal methane gas.

The techniques actually employed include the measurement of the degree of darkening of fossil pollen grains or color changes in conodont fossils, both diagnostic of the thermal alteration of the sediment, and other geochemical evaluations of mineralogical changes induced by fluctuating paleotemperatures. In general, there appears to be a progressive evolution of crude oil characteristics, depending upon geologic age, from younger, heavier, darker, and more aromatic crudes to older, lighter, paler, more paraffinic types. There are, however, numerous exceptions to his rule, especially in geologically young, tectonically active regions with high geothermal gradients.

The Occurrence of Sulfur in Crude Oil

Sulfur is the third most abundant atomic constituent of crude oils, after carbon and hydrogen. Organic sulfur compounds are present in all known crude oils, but their concentrations can vary widely. Generally, higher density crude oils contain the most sulfur. The total sulfur in crude oil varies from below 0.05 percent (by weight), as in some Pennsylvania oils, to about 2 percent for average Middle East crudes and up to 5 percent or more in heavy Mexican and Mississippi oils.[4] In the United States, crudes are, in general, classified as low-sulfur or high-sulfur oils on the basis of the amount of sulfur contained relative to density. Very heavy oil can have almost 1 percent sulfur and still be considered a low-sulfur oil, but lighter oils require a sulfur content below 0.25 percent to be considered low-sulfur.[5]

In the low and medium molecular ranges, sulfur is associated only with carbon and hydrogen, but in the heavier oil fractions it is frequently incorporated in large polycyclic molecules composed of nitrogen, sulfur, and oxygen. Most of the sulfur in crude oils is thought to have been introduced secondarily, especially in the high-sulfur crudes. The secondary process is a chemical reaction between the oil and sulfur or hydrogen sulfide to produce sulfur-containing organic compounds. Thus, although some sulfur-containing compounds may have been carried over directly from the living organic materials from which the oil was derived, most of the intitial sulfur was introduced into primitive oils by the reaction of oil or protopetroleum with elemental sulfur or hydrogen sulfide present in the sediments. The sulfur compounds accumulate with the hydrocarbon compounds in reservoirs to become an immature oil. This process takes place primarily during the early formation of oil, but perhaps also during the migration of oil from source rocks to reservoir rocks. With increasing geological age and depth of burial, sulfur is removed preferentially from oils.[6] The end product of this process is a mature, lower-sulfur oil; however, at shallow depths or along unconformities the desulfurization of the maturation process may be masked by alteration processes during which sulfur compounds are added to the oil by the reaction of the crude with native sulfur and hydrogen sulfide. As a result, crude oils with an intermediate sulfur compound distribution may be formed. A different form of alteration can also take place when hydrogen sulfide gas interacts with oil to yield very corrosive, mercaptan-rich, high-gravity oils and condensates.[7]

Although all crude oils contain sulfur compounds, crude oils rich in sulfur are more common in carbonate reservoirs than in clastic reservoirs. In carbonate reservoirs, sulfur content may reach 5 percent by weight, a volume seldom attained in oils from clastic reservoirs. Massive incorporation of sulfur into sediments in confined environments of the carbonate-evaporite type of sedimentation and subsequent incorporation of the sulfur into organic matter during oil formation probably caused the high-sulfur oils generated in the Middle East. On the other hand, the reducing environments associated with clastic sedimentation result in lower sulfur crudes. High-sulfur crudes typically occur in the arid zones located in the northern and southern tropical paleoclimatic belts, where carbonate-evaporite sedimentation was frequent.[8]

Inorganic Methane

Some terrestrial methane may have been produced by purely inorganic processes. Methane gas, an important constituent in the atmospheres of several planets, is possibly formed by outgassing from hydrocarbons similar to those in carbon-containing meteorites. These meteorites may now exist deep in the earth, having been incorporated as it was being formed. The vast majority of commercial petroleum deposits, however, appear to have been derived from organic material. The biological generation of petroleum can be duplicated in the laboratory, and various structured hydrocarbons thought to be derivatives of the chlorophyll or hemoglobin of once-living plants and animals have been identified in crude oil. Significant commercial accumulations of inorganic methane have yet to be demonstrated.

The Migration of Petroleum

Accumulations of petroleum are usually found in relatively coarse-grained, porous, and permeable sedimentary reservoir rocks that contain very little, if any, insoluble organic matter. As it is not likely that the huge quantities of petroleum now located in these rocks could have been generated from organic material of which no trace remains, the place of origin of petroleum apparently is not always identical to the location in which it is finally found in commercial quantities.

Fluid petroleum compounds are believed to have been generated in significant volumes only in fine-grained sedimentary rocks (usually clays, shales, or clastic carbonates) by geothermal action on kerogen. An insoluble organic residue normally remains in the source rock, at least through the oil-generating stage. The release of fluid petroleum from the solid particles of organic kerogen in the fine-grained source beds and its movement within and through the narrow pores and capillaries of the bed is called primary migration (see Figure 1.1). The petroleum expelled from a source bed then moves through the wider pores of more permeable, coarser-grained rocks (such as sandstones and carbonates). This movement is termed secondary migration. Oil and gas may migrate through such permeable carrier beds until it is trapped by an impermeable or very low permeability barrier, forming a petroleum accumulation, or it may continue its migration until it becomes a seep at the surface of the earth, where it will be broken down chemically by oxidation and bacterial action.

As almost all the pores in the subsurface sediments of sedimentary basins are water saturated, the migration of petroleum compounds takes place in an aqueous environment. Secondary migration may result from active water flow or can occur independently, by either displacement or diffusion. The distinction between primary and secondary migration is based on pore size and rock type. Because the specific gravity of the saline formation water in the larger pores of carrier beds and reservoir

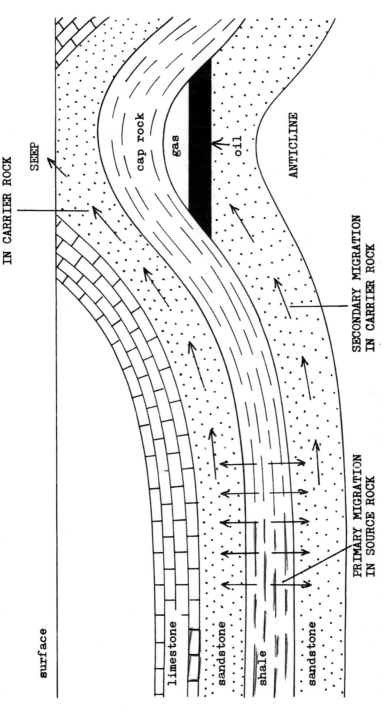

Figure 1.1. Primary migration in shale, secondary migration in sandstone, and entrapment of oil and gas in an anticlinal trap. Note that the petroleum expelled into the upper sandstone is lost to the surface as a seep.

rocks (primarily sandstones and carbonates) is considerably higher than that of oil and gas, in the course of geologic time the petroleum will float to the surface of the formation water and accumulate in the highest portions of any existing traps. The traps are structural highs where reservoir rocks are sealed by relatively impermeable cap rocks, such as shales or evaporites.

The increasing pressure of accumulating sediments also can provide energy to the migration system. The lowermost layers of accumulating sediments experience increasing pressure and temperature related to depth of burial and geothermal gradient. The primary migration of petroleum from a kerogen-containing clay source bed may be initiated during compaction due to the pressure of overlying sediments. After oil generation has begun, continued burial causes clay to become dehydrated by the removal of water molecules that were loosely combined with the clay minerals. Continued temperature increases could result in the newly formed hydrocarbons becoming sufficiently mobile to leave the source beds in solution, suspension, or emulsion, with the water being expelled from the compacting molecular lattices of the clay minerals. The petroleum compounds would comprise only a very small portion of these migrating fluids—a few hundred parts per million.[9]

The general direction of these fluids, moving from a primary to a secondary migration mode, would generally be outward and upward, from regions of higher to lower geostatic pressures. However, in the compacting sediments the migrating fluids would move along complex paths, making lateral or even downward movement locally possible.

The distances over which secondary migration operates are uncertain. Some source rocks are also reservoir rocks and there are many examples of chemically distinct crude oils occupying reservoirs that are separated by only thin zones of impermeable rock. At the other extreme, long-distance secondary migration must have occurred to account for the huge volumes of petroleum that have accumulated in many of the giant fields. It is possible that, under favorable circumstances, extensive carriers exist in the form of permeable beds, unconformities, or bedding planes that have served as conduits for migrating petroleum, which moved for tens or even hundreds of miles through rock strata that was mostly lithified. It is difficult to identify such carriers, as petroleum moving in large volumes of water may not leave distinct traces.

Primary migration probably removes only a small portion of the hydrocarbons that have been formed. The remainder are disseminated as bituminous material or kerogen. Fine-grained carbonate muds can also act as source rocks for petroleum. These muds are not compacted to the same degree as clays, nor is there a mineralogical change that could act as an expulsive process for the contained hydrocarbons. This may account for the fact that many carbonate source rocks are also reservoirs, with no primary migration having occurred. However, if adequate permeability exists, secondary migration due to buoyancy separation would encourage the formation of gas, oil, and water layers.

The Accumulation of Petroleum

The porosity (volume of pore spaces) and permeability (capacity for transmitting fluids) of carrier and reservoir beds are important factors in the migration and accumulation of petroleum. The porosity and permeability of sedimentary rocks may be either primary or secondary in origin. Primary porosity and permeability are dependent upon the size and shape and the grading and packing of the rock particles that constitute the sediment and the manner of their initial consolidation. Secondary porosity and permeability result from post-depositional factors such as solution, recrystallization, fracturing, weathering during temporary exposure at the surface, and further cementation. These factors may either enhance or diminish primary porosity and permeability.

Petroleum is ultimately collected through secondary migration in porous and permeable reservoir rock by interdiction by a trap. The fundamental characteristic of a trap is an upward convex form of permeable reservoir rock, sealed above by a more dense, relatively impermeable cap rock. The convexity may take any shape, the critical factor being that the trap is closed in the form of an inverted container.

A rare exception is hydrodynamic trapping, in which high water saturation of low permeability sediments reduces hydrocarbon permeability to near zero, resulting in a water block and an accumulation of gas or oil down the structural dip of a sedimentary bed, below the formation water.

Structural Traps

Petroleum traps are formed in many ways. Those formed by tectonic events such as the folding or faulting of rock units are called structural traps. The most common structural traps are anticlines (see Figure 1.1), upfolds of rock appearing as ovals on the horizontal planes of geological maps. About 80 percent of the world's oil and gas has been found in anticlinal traps.[10] Although most anticlines are produced by lateral pressure, some have resulted from the draping and subsequent compaction of accumulating sediments over an existing high area, such as a granite ridge or a reef.

The closure of an anticline is the vertical distance between its highest point and the spill plane, the level at which the petroleum could escape if the trap were filled to its maximum capacity. Some traps are filled to their spill plane and others contain considerably smaller amounts of petroleum than their dimensions could accommodate. The petroleum content of a given structure is indicated by the height of the oil and gas column, the vertical distance between the oil/water contact plane and the highest point of the trap at which petroleum is found.

Regional uplifts or arches, several hundred miles long, are found in many sedimentary basins. Petroleum accumulations in these features, however, usually are limited to local crestal closures along the larger structures. Most oil- and gas-producing anticlines are only a few miles long and wide, although those containing giant fields are often much larger.

Fault traps are structural traps in which rock fracture results in relative displacement of strata in a manner that causes a barrier to petroleum migration. This barrier can occur when an impermeable bed comes into contact with a reservoir formation (see Figure 1.2). Sometimes the faults themselves provide a seal against the updip migration of petroleum when they contain impervious clay gouge material between their walls. Faults and folds often combine to produce traps, each providing a part of the container for the enclosed petroleum. Faults can also allow partial or complete escape of petroleum from an existing trap if they breach the cap rock seal.

Other structural traps are associated with salt domes (see Figure 1.3). Petroleum accumulations in traps caused by upward movement of salt masses from deeply buried evaporite beds are common in several of the world's sedimentary basins. These traps have formed along the folded or faulted flanks of the salt plug or on top of the plug where the overlying sediments are also folded and faulted.

Stratigraphic Traps

The formation of stratigraphic traps is related to sediment deposition or erosion, in contrast to the formation of structural traps by tectonic events. A stratigraphic trap is bounded on one or more sides by zones of low permeability. However, because tectonics ultimately control deposition and erosion, few stratigraphic traps are completely devoid of structural influence. The geological history of most sedimentary basins contains the prerequisites for the formation of stratigraphic traps. Typical examples of these traps are fossil carbonate reefs, marine bar sandstone bodies, and deltaic distributary channel sandstones. When buried, each of these geomorphic features provides a permeable, potential reservoir often surrounded by finer-grained sediments that may act as source or cap rocks.

Sediments eroded from a land mass and deposited in an adjacent sea range in grade from coarse- to fine-grained with increasing depth

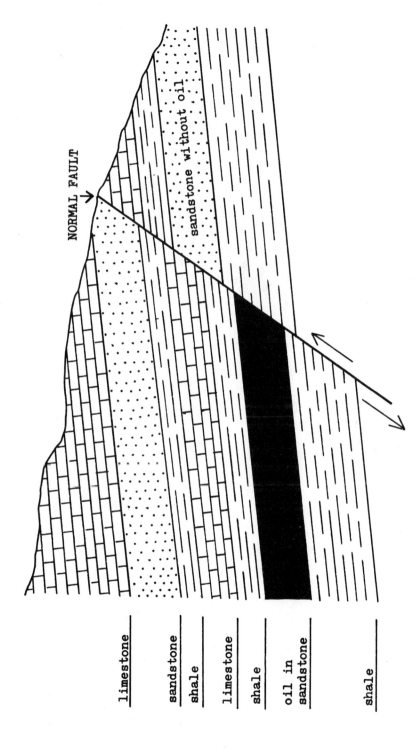

Figure 1.2. Fault trap in which fault movement brought impermeable shale into contact with sandstone reservoir.

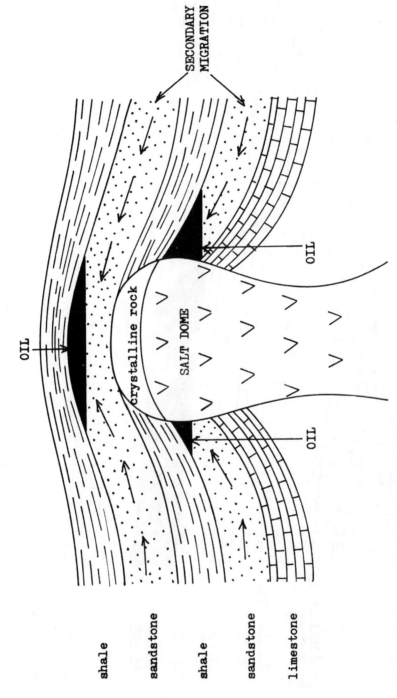

Figure 1.3. Petroleum accumulations in traps caused by upward movement of salt dome.

of water and distance from shore. Permeable coarse-grained deposits thus grade into impermeable fine-grained deposits of the same age. After lithification, tilting may result in the accumulation of any petroleum, which may be migrating through the coarser portions of the bed, against the permeability barrier created by the deposition of the finer-grained material. Such gradation of sediments of one type to another in beds of essentially the same age is called a facies change. Shoreline facies changes from relatively coarse sands to impermeable shales and clays can produce significant petroleum accumulations. In compact carbonate rocks, porosity and permeability may be induced by the dolomitization (substitution of magnesium carbonate for a portion of the original calcium carbonate) of the existing limestones. The updip limit of the dolomitization process (the dolomite facies) will therefore coincide with the limit of petroleum accumulation against the permeability barrier of the dense, unaltered limestone.

Other stratigraphic traps are the result of a break in the sequence of local sedimentary deposition. These breaks are known as unconformities and are produced by uplift and erosion, which sometimes greatly increase porosity and permeability because of groundwater circulation, weathering, and the recrystallization of surface rocks. If an emergent area is resubmerged and such cycles are common over geologic time, and if impermeable beds are deposited over the existing beds with secondary porosity, a potential barrier to petroleum migration may be formed.

Marine transgressions (seas advancing shoreward in response to rising sea level and/or tectonic movements) over previously exposed and weathered sediments have produced many important petroleum accumulations by the deposition of impermeable beds that capped the updip edges of carrier beds. When identified in the subsurface, the underlying permeable beds are said to "wedge out" against the overlying cap rock, with the oil accumulations located below the unconformity. A related type of stratigraphic trap is formed when permeable sandstones are deposited during a marine regression (fall in sea level) and subsequently sealed by the deposition of overlying shales. In this case, the traps, once located in the subsurface, would be above the unconformity, and the underlying permeable beds would act as carriers of petroleum.

Barren Traps

Most petroleum accumulations have been found in clastic reservoir rocks (sandstones and siltstones). Next in number are the carbonate (limestone and dolomite) reservoirs. Petroleum accumulations also occur in the fracture porosity of shales and igneous and metamorphic rocks, but these kinds of deposits are relatively rare. Porosities in reservoir rocks usually range from about 5 to 30 percent, but traps are never full in the sense that all available pore space is occupied by petroleum. A certain amount of residual formation water, which cannot be displaced by migrating petroleum, always remains.

A trap containing only saline formation water is considered barren. The absence of petroleum may be due to inadequate source rock or reservoir rock, insufficient geothermal heat or geologic time for petroleum generation, an inadequate cap rock, or the dispersal or destruction of the accumulated petroleum. Since both source rocks and reservoir rocks are relatively common in sedimentary basins, other causes, such as the interruption of the sequence of events necessary for the formation of a petroleum deposit, commonly account for barren traps. The dispersal or destruction of petroleum accumulations may occur as a consequence of intense tectonic activity, such as significant folding or faulting; igneous activity or metamorphism; the tilting of a trap, allowing the contained petroleum to escape; or uplift and erosion, which may expose the petroleum in a reservoir to degradation by oxidation and bacterial action.

A reservoir need not be completely exposed by surface erosion for its petroleum to be lost. The destructive activity of bacteria, transported downward by circulating groundwater, has probably destroyed many oil deposits.[11] Apparently a thick sedimentary cover is necessary to provide the increased temperature needed for oil generation and to protect a petroleum accumulation from degradation by water-borne surface bacteria and from the fracturing of thin cap rock by tectonic movements. Thinner sealing beds sometimes can be effective caps, especially if they are sufficiently plastic to bend rather than break when subjected to tectonic forces. Evaporites are particularly effective as cap rocks because of their plasticity and impermeability.

Giant Petroleum Accumulations

Most of the world's petroleum is found in large fields, but most fields are small. This is true for sedimentary basins, for countries, and for the world. Since the exploration for petroleum began, approximately 30,000 fields have been discovered. More than 90 percent of these fields are insignificant in terms of their impact on world petroleum production. The two largest classes of fields are *supergiants*—fields with 5 billion or more barrels of recoverable oil or oil-equivalent (oil plus gas) or 30 trillion cubic feet or more of recoverable gas—and *world-class giants*—fields with 500 million to 5 billion barrels of recoverable oil or oil-equivalent or 3 to 30 trillion cubic feet of recoverable gas. The remaining class is used only in the United States and thus can be termed *national-*

class giants—fields with 100 million to 500 million barrels of recoverable oil or oil-equivalent or 0.6 to 3 trillion cubic feet of recoverable gas.

To date, only 37 super-giant oil fields have been found (see Table 11.1), but these few fields originally contained 51 percent of all the oil thus far discovered in the world. The Arabian-Iranian basin in the Persian Gulf region contains 26 of these super-giant fields. The remaining super-giants are distributed as follows: 2 in the United States (North Slope of Alaska and East Texas), 2 in the Soviet Union (West Siberia and Volga-Urals regions), 2 in Mexico (Reforma and Campeche areas), 2 in Libya, 1 in Algeria, 1 in Venezuela, and 1 in China.[12]

The nearly 300 world-class giant oil fields thus far discovered, when added to the 37 super-giants, account for almost 80 percent of the world's known recoverable oil. There are, additionally, approximately 1,000 known relatively large oil fields that contain between 50 and 500 million barrels of oil and account for 14 to 16 percent of the world's known oil. Therefore, less than 5 percent of the known fields originally contained about 95 percent of the world's known conventional oil.[13]

A similar situation exists for natural gas. Free gas in world-class giant and super-giant gas fields and associated gas in some of the world-class giant and super-giant oil fields represent over 80 percent of the world's proved and produced natural gas reserves. The amount of natural gas discovered in the world to date amounts to only about one-third of the oil and natural gas liquids discovered (on a Btu or barrels-of-oil equivalent basis). Since 1965, however, there has been an increase in the discovery of giant gas reserves, while during the same period the discovery of giant oil reserves has declined. An analysis of giant gas fields and their comparison to giant oil fields on the basis of geologic characteristics and sedimentary basin relationships suggests that the world's ultimate resource of natural gas will approach its ultimate resource of oil.[14]

Giant petroleum fields and significant petroleum-producing basins are closely associated. The 7 most productive basins contain 33 of the 37 super-giant oil fields. The Arabian-Iranian basin dominates world oil because of its 26 super-giant fields. (Only 2 other basins have as many as 2 such fields.) Of the major producing basins, only the Permian basin in the southwestern United States does not have a super-giant field. It does have a set of related fields on the eastern edge of its central basin platform whose combined reserves add up to super-giant size. The relative field sizes within a petroleum basin appear to depend upon whether a significant regional arch was developed along with the basin. If a regional arch is not present, smaller structural and stratigraphic traps would then become the sites of accumulation, with the largest field containing less than 10 percent of the basin's total reserves.[15]

In 20 of the 26 most significant oil-containing basins, the 10 largest fields originally contained over 50 percent of known recoverable liquid petroleum reserves. In 9 of these basins more than 75 percent of known reserves were in the 10 largest fields. Only 2 of the 26 basins had less than 30 percent of known reserves in their 10 largest fields. These were both late Tertiary delta-type basins, with no regional arching.[16] Known conventional world petroleum reserves are concentrated in a few sedimentary basins in a relatively small number of giant fields.

Petroleum-Containing Basins

A discussion of the earth's sedimentary basins and subsequent discussions of the petroleum prospects of several regions necessitate references to rocks of various geological ages. The major time divisions of the earth are shown in Table 1.1.

The continents constitute about 30 percent of the earth's surface area. They are composed of sedimentary rocks of varying thickness, averaging about 10,000 feet with a maximum of about 50,000 feet in the deep sedimentary basins. The sedimentary rocks are underlain by about 65,000 feet of granites and metamorphic gneisses that outcrop at the surface in some regions. Below this layer is a denser basaltic layer, thought to be composed of magmatic eclogites and peridotites, which does not reach the surface. This layer is differentiated from the overlying granites and gneisses by a change in seismic velocity, termed the Mohorovicic seismic discontinuity. The oceans are underlain by a relatively uniform basalt crust only about 40,000 feet thick, above which is a thin cover of mainly abyssal sediments. The Mohorovicic discontinuity occurs at shallower depths beneath the oceans than beneath the continents, and the denser magmatic materials occur directly below the relatively thin oceanic basalt layer.

About 25 percent of continent area consists of ancient, highly metamorphosed, Precambrian rocks that seldom, if ever, have been covered by the seas. These areas are called shields and form the stable nuclei of the continents. At various times during their geologic history, the shields have been tilted and their edges submerged. As the shallow seas encroached, sediments were deposited around the central parts of the shields. The thickness of these sediments does not usually exceed 15,000 feet. Much of the earth's land surface is composed of the resulting gently dipping, low-relief regions underlain by these rocks. These regions are known as platforms or shelf areas and, together with the Precambrian shields, make up the central stable portions of the continents. The

TABLE 1.1   Major Time Divisions of the Earth

Era	Period	Epoch	Age (millions of years ago)
Cenozoic	Quaternary	Recent	
			0.01
		Pleistocene	
			2
	Tertiary	Pliocene	
			5
		Miocene	
			25
		Oligocene	
			38
		Eocene	
			56
		Paleocene	
			65
Mesozoic	Cretaceous		
			136
	Jurassic		
			190
	Triassic		
			225
Paleozoic	Permian		
			280
	Pennsylvanian		
			320
	Mississippian		
			345
	Devonian		
			395
	Silurian		
			435
	Ordovician		
			500
	Cambrian		
			570
Precambrian			
		(+ or −)	4,500

Note: The ages mark the boundaries of the time divisions.

shields and platforms together are called cratons. Some regions of the shelves are less stable than the central cratons. These less-stable regions form the transition or hinge-line zones between the cratons and their mobile margins. The hinge lines are subject to cyclic sedimentation. The mobile margins of the cratons are affected by tectonic forces that create a series of downwarps, which may continue to subside for long periods of geologic time. Great thicknesses of sediments accumulate in such basins (see Figure 1.4).

On the basis of geophysical evidence, the earth's crust can be divided into a number of large lithospheric plates whose boundaries are marked by seismic and volcanic activity (see Figure 1.5). The comprehension

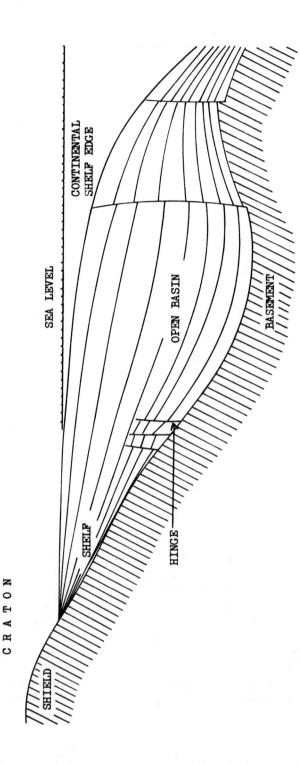

Figure 1.4. Diagram of an open basin showing thinner shelf sediments grading into thick basin sediments on the continental margin.

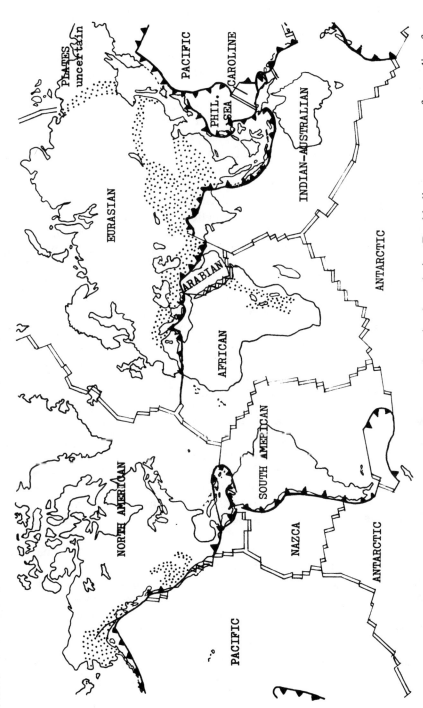

Figure 1.5. Lithospheric plates of the world showing presently active boundaries. Double lines are zones of spreading from which plates are moving apart. Lines with barbs are subduction zones with the barbs on the overriding plate. Single lines are strike-slip faults along which plates slide past each other. Stippled areas are parts of continents, not including plate boundaries, that are tectonically active. Simplified in complex areas.

of plate tectonics requires a global view. The atmosphere moves around the earth at rates measured in tens of miles per hour, while the great ocean currents move at rates of several miles per hour, both in response to the earth's gravity, spin, and heat distribution. In this context, a movement of several inches per year in the "solid" earth may appear to be a reasonable hypothesis, especially on the global scale, in which the earth apparently behaves as a viscous fluid. The concept of plate tectonics requires such movement, for according to this hypothesis, the earth's outer shell (or lithosphere), which is about 60 miles thick, floats on a viscous material (called the asthenosphere) that allows divergences and convergences between adjacent lithospheric plates.

The lithosphere is thought to pull apart in some regions, forming rifts (such as those in East Africa and along mid-ocean ridges), and to converge in other areas to form subduction zones (such as in Japan, the west coast of South America, and Indonesia). The lithosphere is thus a mosaic of semirigid plates that move relative to each other, diverging, converging, and otherwise slipping parallel or oblique to their boundaries. The Himalaya Mountains are thought to have resulted from the collision of the continental plates bearing India and Asia. The relative motion of lithospheric plates provides the basis for the classification of plate boundaries, and the distribution of earthquakes suggests the location of such boundaries and outlines the mosaic-like fit of the various plates. Plate boundaries can be classified as divergent, convergent, or transform. These classes are represented by extension faults, compression faults, and strike-slip faults.

The three different types of continental margins, based on the movements of the lithospheric plates, exhibit distinctive physiography. Divergent margins (extension faults) display prominent continental rises, gentle continental slopes, broad continental shelves, and coastal plains. In contrast, deep oceanic trenches, steep continental slopes, narrow shelves, and island ridges or mountainous uplifts characterize convergent margins (compression faults). Transform margins (strike-slip faults) exhibit peninsula and island separations and borderlands of subsided continental fragments.

New oceanic crust forms at oceanic ridges in a process called seafloor spreading. Here plates apparently diverge along spreading trends, resulting in the upwelling of magma from the asthenosphere to form new oceanic crust. This void-filling process results in a differentiated sequence of rocks, called the ophiolite sequence, usually several miles thick, which includes gabbro at the base, dike swarms above the gabbro, and pillow basalts above the dike swarms.[17]

The formation of new oceanic lithosphere along oceanic ridges implies the destruction of lithosphere elsewhere, apparently in the subduction zones at converging plate boundaries. It has been estimated that converging rates have occurred up to several inches per year. The effects of subduction include the formation of a deep oceanic trench, the steepening of the continental slope, the uplift of deep-water sediments in an island arc or mountain trend adjacent to the continental slope, and the formation of a linear trend of volcanoes on the continental side of the uplifted sediments. The isostatic gravity lows that exist along the continental slopes suggest the downwarp of oceanic crust, and the distribution of shallow, intermediate, and deep-focus earthquakes along an inclined plane landward from the trench suggests reverse faulting.

Thus, the basic features of the island arc system are: a fore-deep or trench, which is a sea-bottom depression where the seismic shear (reverse fault or "Benioff zone") that dips beneath the islands comes to the surface; a nonvolcanic outer arc or frontal arc, which is a structurally complex ridge sometimes emergent above sea level; an inter-deep, often filled with a substantial amount of sediments; a volcanic inner-arc, usually a mountain range; and a back-arc or back-deep, often filled with a thick sedimentary section. The entire belt is mobile and seismically active.

In summary, according to the plate tectonic hypothesis, ocean floors spread and the crust thus created returns to the mantle at subduction zones. Also, the hinge lines of these zones migrate into the subducting slabs, forming new crust behind the subduction zone. Subduction can start, stop, or reverse its direction. Island arcs are formed in regions of subduction. The motions of very large plates can continue for many millions of years, but the detailed interactions by which these large-scale motions are accommodated change repeatedly and relatively rapidly.

The mechanism of plate tectonics is difficult to explain. None of the mechanisms proposed thus far can be reconciled with the geometry and behavior of plate boundaries, particularly if the hypothesis includes global migration of the large plates. Subduction must occur by gravitational sinking, but such sinking cannot pull plates halfway around the earth. The lithospheric slabs descending in subduction zones cause relatively little attenuation of seismic waves. The slabs appear to be under tension in intermediate depths and under compression below about 200 miles. They appear to be sinking of their own weight. Most descending slabs appear to be continuous, and it is generally assumed that they do not penetrate the discontinuity that occurs at a depth of about 400 miles. A driving force may be provided by the earth's spin. Plate motions may represent slight differences in direction and velocity of rotation around the earth's axis, with velocities being higher in low latitudes than in high ones.[18]

The sedimentary basins that contain the world's oil have been classified according to their origin and location in relation to the major structural features of the earth and the concepts of plate tectonics. Worldwide, some 600 basins and sub-basins are known to occur. About 160 of these basins have yielded petroleum, but only 26 are significant producers and 7 of these account for over 65 percent of total discovered petroleum. Some 240 basins have been explored to some extent without yielding commercial petroleum discoveries. The remaining 200 basins have had little or no exploration. They are not completely unknown, however, as sufficient geological work has been done to indicate their dimensions, amount and type of sediment, and gross structural aspects. Most of these underexplored (or frontier) basins are located in very difficult physical environments, such as in polar regions or under relatively deep waters offshore, or have been restricted from exploration for political reasons.

Basin classifications are useful in appraising the petroleum potential of frontier basins and in estimating the undiscovered petroleum remaining in basins in which production is taking place. In an assessment of the ultimate petroleum potential of a basin, the basin is considered a closed container for petroleum and the processes involved in the formation and evolution of petroleum accumulations are defined. A number of basin classifications have been proposed, most involving six to eight basin types. The following is an adaptation and modification of the classification evolved by Klemme during the past decade.[19] Some aspects of the research of Bally are also incorporated.[20] There is general agreement on the classification of about three-quarters of the world's basins. Opinions differ on the remaining quarter, which are poorly known or controversial due to an evolution that includes a change in type through geological time.

Interior Cratonic Basins

Interior cratonic basins are formed in the interior of cratons by local downwarping. Usually found near Precambrian shield areas, they are flat, saucer-shaped basins that have undergone a single cycle of sedimentation, with clastics slightly more prevalent than carbonates. The age of the sediments is generally Paleozoic and the deposition is of the platform type. The petroleum traps are primarily associated with basement-controlled, regional, central arches, with stratigraphic traps (such as carbonate reefs) occurring around the margins. Interior cratonic basins are not prolific producers of petroleum and contain no known world-class giant fields. About 2 percent of the world's oil and less

TABLE 1.2   Producing Interior Cratonic Basins

Basin	Location	World Class Giant Fields	Yield per Sediment Volume
Illinois	United States	No	Average
Williston	United States (Dakotas) & Canada	No	Low
Denver	United States	No	
Michigan	United States	No	Low
Paris	France	No	Low
Baltic	Northern Europe	No	
Natuna	Indonesia	No	

than 1 percent of the world's gas have been found in interior cratonic basins, yet they make up 18 percent of the world's basin area. Their shallow depths and low geothermal heat flow inhibit the formation of thermal gas. Average recovery for such basins is below 20,000 barrels of oil and equivalent gas per cubic mile of sediments. The best recovery recorded thus far is 35,000 barrels of oil per cubic mile of sediments.

There is continuing exploratory activity in this type of basin in north-central Africa, northern Canada, and northern Europe. Because interior cratonic basins have poor petroleum discovery rates, frontier basins of this type have a 50 percent or less chance of commercial discovery and show little likelihood of containing giant fields.

Examples of producing interior cratonic basins are given in Table 1.2.

Intracontinental Composite and Complex Basins

Intracontinental composite basins occur outward from interior cratonic basins, near the margins of cratons. They are usually shallow but asymmetrical in profile. In general, they were initially sites of Paleozoic platform deposits derived from Precambrian shields, similar to interior cratonic basins, but then became multicycle (or composite) basins when a tectonic uplift on the exterior margin of the craton provided a second cycle of sediments from an opposite source area. The asymmetry of the basin profile is created by the two-sided source of sediments. The second cycle is usually late Paleozoic or Mesozoic in age. Downwarping appears to have occurred by extension of the crust during the first cycle, followed by compression during the second cycle. Basin development may have been related to lithospheric plate movement near the margins of the craton. Intracontinental composite basins have developed around the North American Precambrian shield, west of the

South American shield, and east of the European Fennoscandia and Asian Siberian shields. Composite basins are quite productive in petroleum and many contain giant fields.

Intracontinental complex basins are also multicycle basins located toward the margins of cratons. They are large, usually elliptical basins with an irregular or asymmetrical profile. Their genesis appears to have been more complex than that of the composite basins, with multiple rift faulting followed by rather symmetrical downwarping, similar to that of interior cratonic basins. Complex basins are also very productive and often contain giant fields.

Both types of intracontinental basins contain clastic and carbonate sediments, with petroleum reservoirs being predominantly in clastics. Petroleum traps are usually associated with large arches or block uplifts. In many composite basins a regional arch accounts for up to half the known petroleum in one series of related accumulations, although compression folds and stratigraphic facies changes can also act as traps. Commercial petroleum accumulations occur in sediments deposited during both the lower and upper cycles. Geothermal gradients in intracontinental basins are normal to somewhat above normal (in complex basins), but there is considerable shallow petroleum production, resulting from either post-maturation uplift or secondary migration updip through carrier beds.

The intracontinental basins make up about 27 percent of the world's basin area and contain about one-quarter of the world's known oil and gas. The oil, about 14 percent of the world total, occurs mostly in composite basins, while the gas, 48 percent of the world total, mainly occurs in complex basins. Petroleum recoveries are about average, the yields in the better intracontinental basins being above 90,000 barrels of oil and equivalent gas per cubic mile of sediments (the best recovery to date is 250,000 barrels per cubic mile). Average productive basins yield between 60,000 and 90,000 barrels per cubic mile and poorer productive basins yield below 20,000 barrels per cubic mile. Since about three-quarters of intracontinental basins appear productive and two in three may contain giant fields, frontier basins of this type are considered to have a better than 50 percent chance of commercial discovery.

Examples of producing intracontinental composite and complex basins are given in Tables 1.3 and 1.4.

Rift Basins

Rift basins are small, linear basins of primarily late Paleozoic, Mesozoic, and Tertiary ages that are formed in down-faulted grabens on or near cratonic areas. A majority are developed along the trends of older tectonic belts in divergent regions. The basins are irregular in

TABLE 1.3   Producing Intracontinental Composite Basins

Basin	Location	World Class Giant Fields	Yield per Sediment Volume
Big Horn	United States (Wyoming)	Yes	Average
Powder River	United States (Wyoming)	Yes	
Wind River	United States (Wyoming)	No	Average
Green River	United States (Wyoming)	No	Low
Paradox	United States (Utah)	No	
Uinta	United States (Utah)	No	
Piceance	United States (Colorado)	Yes	
San Juan	United States (New Mexico)	No	
Permian	United States (Texas)	Yes	High
Anadarko	United States (Oklahoma)	Yes	
Arkoma	United States (Oklahoma)	No	Low
Appalachian	Eastern United States	Yes	
Alberta	Canada	Yes	Average
Volga-Urals	Soviet Union	Yes	High
Vilyuy-Lena	Soviet Union	Yes	
Sub-Caspian	Soviet Union	No	
Szechwan	China	Yes	
Ordos	China	No	
Aquitane	France	Yes	Low
Oriente	Ecuador	Yes	
Bolivia	Bolivia	No	
Neuquén	Argentina	No	
North Argentina	Argentina	No	
Magallenes	Southern South America	No	Average

profile and filled with mostly clastic sediments, but carbonates also occur in rift basins that were open to warm seas during portions of their geologic history. The genesis of rift basins includes extensional crustal features with considerable block-fault movement, resulting in combination structural-stratigraphic traps where variations in deposition have occurred over differentially subsiding blocks. Secondary migration of petroleum is often only for short distances. Geothermal gradients are normal to high and organic materiral is often over mature.

Rift basins account for only about 5 percent of the world's basins by area, and about half of them are productive, but they contain about 10 percent of the world's known petroleum (12 percent of the oil and 4 percent of the gas). Recovery is good: In rift basins that contain marine or lacustrine sediments, above-average oil recovery per unit volume of sediment has been recorded. In the best basins, recovery

TABLE 1.4   Producing Intracontinental Complex Basins

Basin	Location	World Class Giant Fields	Yield per Sediment Volume
West Siberia	Soviet Union	Yes	High
Pechora	Soviet Union	Yes	
Grand Erg Oriental	North Africa	Yes	
Polignac	Algeria	Yes	
Artesian-Cooper	Australia	Yes	
Surat	Australia	No	
Southern North Sea	North Sea	Yes	
Northwest Germany	Germany	No	

has been as high as 450,000 barrels of oil and equivalent gas per cubic mile of sediments. Average productive rift basins are in the 90,000 to 300,000 barrels-per-cubic-mile recovery range, while the poorer producing basins have been found to yield between 20,000 and 60,000 barrels per cubic mile.

As about half of the explored rift basins have been found to contain commercial petroleum deposits, frontier rift basins may have about a 50 percent chance of commercial discovery.

Examples of producing rift basins are given in Table 1.5.

Extracontinental Closed, Trough, and Open Downwarp Basins

The closed and trough-type extracontinental basins occur at the margins of cratons. The closed extracontinental basins resemble the composite basins of the craton but are formed along the tectonic borderlands of the lithospheric plates. The open basins are present adjacent to small ocean basins, such as the Arctic Ocean, the Gulf of Mexico, the Caribbean, the eastern Mediterranean, and the South China Sea. Their genesis is related to the evolution of the small ocean basins into which they open. The open basins may become closed as a result of the collision of continental lithospheric plates. Upon closing, a large asymmetric basin results, with a two-sided source of sediments, which is similar to an intracontinental composite basin. Further plate movement may destroy a considerable portion of the closed basin, leaving only a downwarped trough. For example, the narrow troughs south of the Himalayas on the Indian-Australian plate were formed by collision with the overriding Eurasian plate. Other such troughs are the North African and narrow Alpine troughs.

TABLE 1.5   Producing Rift Basins

Basin	Location	World Class Giant Fields	Yield per Sediment Volume
Sirte	Libya	Yes	High
Suez	Middle East	Yes	
Viking	Norwegian Sea	Yes	
Irish Sea	Irish Sea	No	
Rhine	Germany	No	Low
Poland		No	
Dnieper-Donetz	Soviet Union	Yes	
Cambay-Bombay	India (offshore)	Yes	
Sungliao	China	Yes	
Bohai	North China	Yes	
Tsaidam	China	Yes	Average
Dzungaria	China	Yes	Average
Gulf of Siam	Thailand (offshore)	Yes	
Reconcavo	Brazil	No	Average
Central African grabens		No	

Extracontinental downwarp basins represent only 18 percent of the world's basin area but are extremely prolific petroleum producers. They account for about 48 percent of the world's known petroleum (54 percent of the world's oil and 38 percent of the world's gas). The most productive of these basins is the relatively unique Arabian-Iranian basin, which is locally submerged by the Persian Gulf.

The large amounts of petroleum that have been found in both closed and open extracontinental downwarp basins may be related to the above normal (often high) geothermal gradients, which may provide efficient petroleum generation along with shallow primary migration and secondary migration over long distances. Often, in contrast to their cratonic counterparts, the downwarp extracontinental basins contain relatively large amounts of rich source shales and evaporites (which are excellent cap rocks). Also, considerable thicknesses of porous carbonates, which provide the predominate reservoir rocks, occur. In closed and trough basins the traps are mostly anticlinal: either sediment drape over large, regional arches or compressional folds associated with salt movement. In the open basins, combination stratigraphic-structural and stratigraphic traps are more common, but salt-induced structures also occur. The ages of the sediments in downwarp basins are late Paleozoic, Mesozoic, and Tertiary.

The historic risk of exploring these basins appears to be about a 50

TABLE 1.6   Producing Extracontinental Closed Downwarp Basins

Basin	Location	World Class Giant Fields	Yield per Sediment Volume
Arabian-Iranian	Middle East	Yes	High
Southern Soviet basins		Yes	
Maturín	Eastern Venezuela	Yes	
North Australia		No	

TABLE 1.7   Producing Extracontinental Trough Downwarp Basins

Basin	Location	World Class Giant Fields	Yield per Sediment Volume
Indus	South Asia	Yes	
Assam-Bangladesh	South Asia	No	
Po Valley	Italy	Yes	
East Italy		No	
Molasse	Central Europe	No	Low
Ploesti	Rumania	Yes	
Potwar	North Africa	No	

percent chance of finding commercial production and about a one in three chance of finding giant fields. In the most prolific of the open and closed downwarp basins recovery is high: over 300,000 barrels of oil or equivalent gas per cubic mile of sediment, with the Arabian-Iranian basin the richest at about 600,000 barrels per cubic mile. For average producing basins recovery is between 90,000 and 300,000 barrels per cubic mile. The poorest of the productive open and closed downwarp basins contain between 60,000 and 90,000 barrels of known oil and equivalent gas per cubic mile of sediment. In the trough basins, recovery is not nearly as good. Most of these basins that are productive contain from 20,000 to 60,000 barrels of known oil and equivalent gas per cubic mile of sediment, but the poorest contain less than 20,000 barrels per cubic mile.

Examples of producing extracontinental closed, trough, and open downwarp basins are given in Tables 1.6, 1.7, and 1.8.

Extracontinental Pull-Apart Basins

Extracontinental pull-apart basins occupy the intermediate zone between the thick continental and the thin oceanic crust. They are located

TABLE 1.8   Producing Extracontinental Open Downwarp Basins

Basin	Location	World Class Giant Fields	Yield per Sediment Volume
Gulf Coast	United States	Yes	High
North Slope	United States (Alaska)	Yes	
Sverdrup	Northern Canada	Yes	
Tampico	Eastern Mexico	Yes	High
Reforma-Campeche	Southern Mexico	Yes	High
North Borneo–Sarawak		Yes	
East Mediterranean		No	
Gulf of Gabes	Tunisia	No	

along the major oceanic boundaries of spreading plates and appear to be initiated by rifting over Precambrian zones of weakness or along older tectonic trends. As the evolution of the basin proceeds, one or more of the rifts becomes the site for the upwelling of oceanic basalt and thus seafloor spreading is established. As the spreading continues, the region subsides and a sedimentary fan is deposited seaward from the adjacent continent.

Pull-apart basins may form parallel or transverse to the axis of seafloor spreading. They are deep linear basins with high volumes of sediments derived from a one-sided source and are, therefore, asymmetrical in profile. Due to their extensional genesis, most traps in these basins are tensional growth anticlines. Geothermal gradients are generally normal to low, when distant from the spreading sites, and oil accumulations, therefore, can be deeper than normal, although the rifts themselves may exhibit high heat flow. The sediments are mostly clastics, but the post-rift sequence may include a carbonate bank along with the clastic fan.

The pull-apart basins are primarily of Mesozoic and Tertiary age and represent about 18 percent of the surface basin area of the world. Only 10 percent of these basins are petroleum productive, mainly because 90 percent of them are located offshore (55 percent in deep water). Pull-apart basins have displayed low productivity to date (containing less than 1 percent of the world's petroleum), but their eventual contribution will greatly affect the magnitude of the world's ultimate recoverable petroleum resources. The best recovery to date in a pull-apart basin is between 20,000 and 60,000 barrels of oil or equivalent gas per cubic mile of sediment. The poorer producing basins have had recoveries below 20,000 barrels per cubic mile. The statistical experience

TABLE 1.9   Producing Extracontinental Pull-Apart Basins

Basin	Location	World Class Giant Fields	Yield per Sediment Volume
Northwestern Australia		Yes	
Perth	Australia	No	Average
Angola-Cabinda	Africa	Yes	Average
Gabon	Africa	No	Average
Ivory Coast–Ghana	Africa	No	
Palawan	Philippines	No	
Campos	Brazil	No	
Sergipe-Alagoas	Brazil	No	Average
Northeast Brazil		No	
Newfoundland	Canada	Yes	

with these basins does not permit much speculation, but the success ratio for significant commercial discoveries has been about one in three. The pull-apart basins off the east coast of the United States are currently being drilled, so far with minor success. As offshore drilling technology improves and the price of oil rises, however, more of these kinds of basins will be explored.

Examples of producing extracontinental pull-apart basins are given in Table 1.9.

Extracontinental Subduction Basins

Subduction basins are small, shallow, linear, borderland basins with irregular profiles that are formed along convergent margins involved in lithospheric plate subduction. They are classified according to their relation to the volcanic island arc often present on the overriding cratonic plate above the subduction zone.

Fore-arc basins are located on the oceanward side of the volcanic arc, along the boundaries of the major crustal plates. They are thought to originate either by the dragging down of the substratum or by the piling up of sediments and subsequent sinking of the crust along a subduction zone. Sediments consist of shales and carbonates, sometimes interbedded with volcanics. Due to low heat flow, poor reservoir rocks, and a lack of trapping structures, these basins have been relatively poor prospects for commercial petroleum deposits.

Back-arc basins (sometimes called foreland basins) are located on the cratonic side of the volcanic arc. These basins are formed by the collapse or tilting of cratonic plates with associated foundering of continental margins. Commonly, they contain prograding wedges of Tertiary clastics derived from the adjacent craton, carbonate platform rocks, and abundant amounts of organic material. Sediments draped over fault blocks provide the main petroleum traps, but compressional anticlines and stratigraphic traps occur in some basins. High geothermal gradients provide for highly efficient maturation and migration of hydrocarbons. The multiple-pay reservoirs are predominantly sandstones, although petroleum is also found in carbonate rocks. Some of these basins have extremely high oil recovery per sediment volume.

Inner-arc basins occur in the same relative tectonic position as back-arc basins, but without an adjoining craton. Therefore, they do not have a source for prograding wedges of sedimentary rocks. Many inner-arc basins contain thin and geochemically immature sedimentary sections, and thus have only marginal petroleum potential. Some of these basins, however, do have relatively thick sections of sedimentary rocks, and the deeper parts of such sections may have been subjected to sufficiently high temperatures to generate oil. As many inner-arc basins are located in relatively deep waters, few have been explored.

Non-arc basins represent a continuation of plate margin convergence until subduction and wrench faulting destroy the island arc. These basins are most often Tertiary in age and are filled with clastic sediments. The main traps are sediments draped over fault blocks and structural anticlines. Maturation and migration of petroleum is very efficient due to high geothermal gradients. Reservoirs are of the multiple-pay type, predominantly sandstones, ranging from shallow marine deposits to turbidites. Oil recovery per sediment volume in the best of the non-arc basins is the highest of all the world's basin types.

Extracontinental subduction basins represent 7 percent of the world's basin area, and only about one-quarter of them (2 percent of the world's basin area) are productive. However, they contain about 7 percent of the world's known petroleum (over 8 percent of the world's oil and 2 percent of the world's gas). Subduction basins commonly have a single field that contains between 25 and 45 percent of the basin's total petroleum. The best of the subduction basins contain more than 300,000 barrels of oil or equivalent gas per cubic mile of contained sediments. The highest yield is 4 million barrels per cubic mile! Average basins contain between 90,000 and 300,000 barrels per cubic mile, while the poorest of the producing subduction basins contain below 20,000 barrels of known oil-equivalent per cubic mile of sediments.

Examples of producing fore-arc, back-arc, inner-arc, and non-arc subduction basins are given in Tables 1.10 through 1.13.

TABLE 1.10   Producing Fore-arc Subduction Basins

Basin	Location	World Class Giant Fields	Yield per Sediment Volume
Talara	Northwest coast South America (Peru)	Yes	
North Ceram	Indonesia	No	

TABLE 1.11   Producing Back-arc Subduction Basins

Basin	Location	World Class Giant Fields	Yield per Sediment Volume
Central Sumatra	Indonesia	Yes	High
North Sumatra	Indonesia	Yes	
South Sumatra	Indonesia	No	
Northwest Java	Indonesia	No	
East Java	Indonesia	No	
Salawati	Indonesia	No	
West Honshu	Japan	No	
Sakhalin	Soviet Union	No	
Cook Inlet	Alaska	Yes	

TABLE 1.12   Potentially Producing Inner-arc Subduction Basins

Basin	Location	World Class Giant Fields	Yield per Sediment Volume
Bone	Indonesia	No	

Extracontinental Median Basins

Median basins are small, shallow, linear basins with an irregular profile that are formed in the folded and intruded "median" zone between either an oceanic subduction zone and the basins of the craton or in the collision zone between two cratonic plates. They are formed by wrench-fault movement and foundering, which creates local tension within the compressed and uplifted mountain belts surrounding the convergent margins of some continents. Geothermal gradients are normal to high. Sediments are predominantly clastic, and petroleum traps are block uplifts over which structural-stratigraphic accumulations occur.

Median basins represent about 3.5 percent of the world's basin area and contain about 2.5 percent of the world's known petroleum. About

TABLE 1.13   Producing Non-arc Subduction Basins

Basin	Location	World Class Giant Fields	Yield per Sediment Volume
Los Angeles	United States (California)	Yes	High
Ventura	United States (California)	Yes	High
San Joaquin	United States (California)	Yes	High
Sacramento	United States (California)	Yes	
Santa Maria	United States (California)	No	High
Salinas	United States (California)	Yes	
Vienna	Europe	Yes	Average
Baku	Soviet Union	Yes	
Crimea	Soviet Union	No	
West Caspian	Soviet Union	No	
Burma		No	

one-quarter of them are petroleum productive, and among these recovery is variable. The best median basins have been found to contain 90,000 to 300,000 barrels of oil-equivalent per cubic mile of sediments, average basins contain between 20,000 and 60,000 barrels per cubic mile, and the poorest have been found to contain less than 20,000 barrels of oil or equivalent gas per cubic mile of contained sediments.

Examples of producing median basins are given in Table 1.14.

Delta Basins

Delta basins are generally deep, small to medium sized, circular, late Tertiary centers of deposition whose clastic sedimentary fill is derived from major continental drainage areas. They contain a large volume of sediments relative to their surface area and are often the site of present-day deltas, which are prograding seaward. Delta basins appear to have developed in many different tectonic settings. About 36 percent have formed over extracontinental downwarp basins, 17 percent along pull-apart basins, 16 percent over subduction basins, 12 percent in rift basins, 12 percent in median basins, and 7 percent over the submerged portions of composite basins. Their location is about equally divided between divergent and convergent lithospheric plate margins. Petroleum traps are primarily tensional structural features and flow structures.

TABLE 1.14   Producing Median Basins

Basin	Location	World Class Giant Fields	Yield per Sediment Volume
Maracaibo	Venezuela	Yes	High
Upper Magdalena	Colombia	Yes	
Gippsland	Australia	Yes	Average
Taranaki	New Zealand	Yes	Low
Kutei	Indonesia	Yes	
Tarakan	Indonesia	No	
Barito	Indonesia	No	
Pannonian	Eastern Europe	No	Low
Central Iran		No	

TABLE 1.15   Producing Delta Basins

Basin	Location	World Class Giant Fields	Yield per Sediment Volume
Mississippi	United States	Yes	
Mackenzie	Canada	Yes	
Niger	Nigeria	Yes	High
Mahakam	Indonesia	Yes	
Nile	Egypt	No	
Po	Italy	No	
North Borneo		No	

Low geothermal gradients result in below-average depths of petroleum accumulation.

Delta basins have a unique field size distribution, involving few giant fields. In deltas of divergent margins, the largest field usually contains less than 7 percent of the basin's petroleum. The deltas of convergent margins, which have inherited the underlying structure, contain more giant fields. Delta basins have a higher than normal gas content because of the predominance of terrestrial organic source material. These basins are relatively rich in petroleum, averaging between 90,000 and 300,000 barrels of oil or equivalent gas per cubic mile of contained sediments. They represent 2.5 percent of the world's basin area and contain 6 percent of the world's known petroleum. About 40 percent have proved production.

Examples of producing delta basins are given in Table 1.15.

Summary

Worldwide, approximately 600 sedimentary basins are known to exist. About 160 of these have yielded petroleum, but only 26 are significant producers and 7 of these account for over 65 percent of total known petroleum. Exploration has occurred in another 240 basins without yielding commercial discoveries. The remaining 200 basins have had little or no exploration. They have had sufficient geological study, however, to indicate their dimensions, amount and type of sediments, and gross structural aspects. Most of the underexplored (or frontier) basins are located in environments that are difficult to exploit, such as polar regions or submerged continental margins. The large basins (over 200,000 cubic miles of sediments) account for 70 percent of known world petroleum. Future exploration will have to involve the smaller basins as well, and these will generally contain less petroleum.

Basins exhibit a wide range of petroleum recovery. Interior basins, in general, have a history of low recovery. This has also been true for pull-apart basins that have been successfully explored. Because of their offshore location (often in deep water), pull-apart basins have so far been little explored, but they are important to the world's eventual total recovered petroleum. Although they represent 18 percent of the surface basin area of the world, pull-apart basins to date have been found to contain less than 1 percent of the world's proved and produced petroleum. If this continues to be the rule, future world reserve additions could be lower than expected.

Subduction, downwarp, rift, and delta basins, when productive, often provide high recovery yields. Downwarp basins contain almost half of the world's proved and produced petroleum, yet constitute only 18 percent of the world's basin area. The highest-yield (per sediment volume) basins are the subduction basins, but they are small in size. Composite and complex basins generally provide average yields. Taken together, the large downwarp basins and composite and complex basins make up almost half of the world's basin area and contain about three-quarters of the world's known petroleum.

A good commercial basin usually has an average recovery ranging from 100,000 to 200,000 barrels of oil or equivalent gas per cubic mile of contained sediments. Poor (but productive) basins range between 1,000 and 20,000 barrels per cubic mile. Petroleum is found in normal quantities if favorable source, carrier, reservoir, and cap rocks are properly related in space and time. However, super-giant fields and very rich basins are often dependent upon additional geological factors such as regional arches, evaporite beds, unconformities, secondary fracturing, secondary porosity, and high geothermal gradients.

As exploration moves into polar regions, all types of basins will be tested: composite and complex, rift, downwarp, subduction, median, and delta, as well as pull-apart and even interior basins (in Siberia). These polar basins are expected to provide the bulk of the world's new petroleum discoveries.

Statistics suggest that about 40 percent of the 200 frontier basins (about 80) may contain commercial petroleum deposits. About 13 could become significant producers (9 billion or more barrels of recoverable oil and/or equivalent gas), but only 3 might contain over 36 billion barrels of recoverable oil-equivalent. No frontier basin seems likely to rival the Arabian-Iranian basin in the size of its petroleum resource (in excess of 677 billion barrels of recoverable oil and 1,750 trillion cubic feet of gas). Due to the location of these frontier basins in difficult environments, petroleum development operations will be more expensive in terms of materials, energy, manpower, and money than in currently productive basins. Thus, only the larger fields will be economic to exploit, leaving a larger portion of the technically recoverable oil behind in small fields than is necessary in less hostile locations. Known conventional world petroleum is concentrated in a few sedimentary basins in a relatively small number of giant fields. This pattern is not expected to change significantly as exploration continues.

Notes

1. Dudley D. Rice and George E. Claypool, "Generation, Accumulation, and Resource Potential of Biogenic Gas," *American Association of Petroleum Geologists Bulletin* (January 1981), pp. 5–25.

2. H. Douglas Klemme, "Geothermal Gradients, Heat Flow, and Hydrocarbon Recovery," *Petroleum and Global Tectonics* (Princeton and London: Princeton University Press, 1975), p. 260; E. N. Tiratsoo, *Oilfields of the World,* 2nd ed. (Houston, Texas: Gulf Publishing Company, 1976), p. 37; and B. P. Tissot and D. H. Welte, *Petroleum Formation and Occurrence* (New York: Springer-Verlag, 1978), p. 185.

3. Rice and Claypool, "Biogenic Gas," p. 6.

4. G. D. Hobson, *Modern Petroleum Technology* (New York: John Wiley & Sons, 1973), p. 193.

5. W. L. Nelson, "Sulfur Content of U.S. Crude Oils," *Oil and Gas Journal* (August 17, 1970), p. 78.

6. T. Y. Ho, M. A. Rogers, H. V. Drushel, and C. B. Koons, "Evolution of Sulfur Compounds in Crude Oils," *American Association of Petroleum Geologists Bulletin* (November 1974), p. 2345.

7. *Ibid.,* p. 2348.

8. Tissot and Welte, *Petroleum Formation and Occurrence,* p. 399.

9. Tiratsoo, *Oilfields of the World,* p. 42.

10. *Ibid.,* p. 46.

11. *Ibid.,* p. 56.

12. Richard Nehring, "The Outlook for World Oil Resources," *Oil and Gas Journal* (October 27, 1980), pp. 170–175.

13. *Ibid.*

14. H. Douglas Klemme, personal communication.

15. H. Douglas Klemme, "To Find a Giant, Find the Right Basin," *Oil and Gas Journal* (March 15, 1971).

16. Nehring, "Outlook."

17. F. J. Vine and E. M. Moores, "A Model for the Gross Structure, Petrology, and Magnetic Properties of Oceanic Crust," Geological Society of America, Memoir 132, 1972, pp. 195–205.

18. Warren Hamilton, "Tectonics of the Indonesian Region," Geological Survey Professional Paper 1078 (Washington, D.C.: U.S. Government Printing Office) pp. 303–307.

19. H. Douglas Klemme, "Petroleum Basins—Classifications and Characteristics," *Journal of Petroleum Geology* (October 1980), pp. 187–207; H. Douglas Klemme, "The Geology of Future Petroleum Resources," *Revue de l'Institut Français du Petrole* (March-April 1980), pp. 337–349; and H. Douglas Klemme, "Giant Oil Fields Related to Their Geologic Setting: A Possible Guide to Exploration," *Bulletin of Canadian Petroleum Geology* (March 1975), pp. 30–66.

20. A. W. Bally, "A Geodynamic Scenario for Hydrocarbon Occurrences," *Proceedings of the Ninth World Petroleum Congress,* v. 2 (1975), pp. 33–44.

2

Exploration and Production

Exploration Methods

In the past all of the exploration for petroleum accumulations in sedimentary basins was done by the use of surface methods. Current exploration employs mostly subsurface techniques.

Surface Methods

Surface methods include the observation of direct indications of the presence of underlying petroleum, such as oil and gas seeps, and geological and geochemical prospecting. Oil seeps indicate the presence of oil accumulations at depth and in rare instances are large enough to make the drawing off and marketing of the surface oil profitable. Gas seeps are difficult to detect unless they are bubbling through water or mud or are ignited.

Seeps are very important in leading to the discovery of petroleum deposits and in fact have been the means of the discovery of many of the world's major oil fields. If folding and faulting are intense, the seepage may be across the bedding, and the leaking trap can be tapped by a well drilled next to the seep. If the folding is gentle, seepage may be through the reservoir horizon from a field located down the dip many miles away. In this case, the seep is evidence of the presence of petroleum in the formation but does not immediately lead to the discovery of an accumulation.

The field geologist is primarily concerned with the interpretation of land forms and rock outcrops in terms of areal and structural geology. Some structural traps, such as anticlines, may outcrop at the surface in a characteristic shape. Anticlinal outcrops are ovals, with the oldest rocks in the center. Drilling along the center axis of such ovals (or inside circular patterns indicative of domes) has resulted in the discovery of many oil fields. In addition, the knowledge of the general geology and the geologic history of a basin is an aid in the delineation of stratigraphic traps and the interpretation of subsurface data.

Geochemical methods of prospecting for petroleum are based primarily on the detection of microseepages of gases and hydrocarbons of low molecular weight from deeper petroleum accumulations into the near-surface soil. The migration that brings these low-molecular-weight hydrocarbons to the surface can be termed tertiary migration to distinguish it from the primary migration in source beds and secondary migration through carrier beds to reservoirs. Tertiary migration (microseeps) occurs by the upward diffusion of natural gas (usually dissolved in groundwater) caused by the flow of molecules from regions of higher concentration to regions of lower concentration in an effort to establish an equilibrium; the upward effusion of gas through rock fractures, caused by pressure differences between the subsurface and the surface; and the hydrodynamic or chemical potential drive of water containing dissolved low-molecular-weight hydrocarbons vertically through capping shales toward the surface.

The geochemical methods used to detect these microseeps are: hydrochemistry, soil-salt methods, trace-metal analysis, oxidation-reduction potential determinations, carbon-isotope determinations, and carbon-dioxide values. These chemical methods attempt to detect anomalous geochemical conditions in the near surface that may be related to the escape of gases and low-molecular-weight hydrocarbons from petroleum traps. The anomalous geochemical areas are called halo patterns. Thus, the foundation of geochemical prospecting is the location of areas that exhibit chemical halo patterns.

Prospecting for petroleum by geochemical soil analysis has generally been unsuccessful due to improper sampling, inadequate sample analysis, contamination of samples, and the ubiquitous character of natural gas. As a reconnaissance tool used in conjunction with subsurface methods, geochemical prospecting may have merit, particularly in areas of less indurated reservoirs and cap rocks at relatively shallow depths.

Subsurface Methods

Subsurface methods of exploration are used where the bed rock is not adequately exposed. This includes areas with a veneer of soil, alluvium, or glacial drift and regions in which the deeper subsurface structure is not reflected in the near-surface beds due to an angular unconformity. One of the first methods employed to penetrate the surface cover was core drilling. The object is to obtain sufficient stratigraphic elevation data for the delineation of any structural anomalies that may be present. The core drilling rig is a lightweight rotary drill mounted on a truck. The usual procedure is to drill a series of core holes in a line in the direction of the regional dip of the strata. Any reversals in elevation of the marker horizons, encountered by the drill and identified in the cores, could indicate a structure at depth.

Geophysical surveys also are used to penetrate surficial deposits and to provide an indication of deeper structures. Magnetic measurements, often taken by airborne magnetometer surveys, may indicate basement highs over which oil-bearing sediments may be draped. The crystalline rocks of the Precambrian basement or basic igneous rock intrusions in the sedimentary section may be detected due to their magnetic intensity. The depth to basement or presence of igneous rocks affects sediment structure and oil-generating capacity. Magnetic measurements normally need to be supplemented by other geophysical tools to define a precise drilling location.

Gravity measurements are used to determine variations in gravitational attraction produced by rocks of different densities. Gravimeters are sensitive to variations in the sedimentary section that may have structural significance. Areas of high gravity may be upfolds that bring denser, older rocks closer to the surface, or they may reflect differences in the basement rock and have no structural significance. Gravity prospecting methods have been successful in the discovery of piercement-type salt domes because of the marked variation in density between the salt and the country rock. The gravity map would contain an area of low gravitational attraction above the buried salt mass. Igneous intrusions, in contrast, appear as areas of high gravitational attraction. However, as both upfolds and downfolds can create the same kinds of gravity anomalies (depending upon the distribution of rock densities), a knowledge of local stratigraphy and tectonics is necessary to interpret gravity records.

The most elaborate, expensive, and effective geophysical tool is the seismograph. The magnetometer and the gravimeter are usually used as reconnaissance instruments, but the seismometer is employed to determine the structural geology in detail. (Both magnetic and gravity surveys, however, can be of great value in locating the deep basins containing thick sedimentary sequences, which are the primary targets for petroleum exploration.)

Seismic surveying is done by either refraction or reflection techniques. Both methods require shock waves to be generated on the surface. These waves travel downward into the subsurface strata and are refracted and reflected back to the surface by rock strata of differing density. The length of time necessary for a round trip is a reliable indication of the depth of the refracting or reflecting layers. Therefore, it is possible to trace the structural pattern of various layers of rock beneath the surface. The shock (or seismic) waves may be generated by either an explosive charge placed in a shallow hole or heavy weights or vibrators that can be accelerated to strike the ground.

The seismic refraction method was originally designed to locate salt plugs and trace major anticlinal structures in massive limestones. The method is based on the travel time of seismic waves, which move faster in salt and limestone than in associated noncrystalline rocks. Seismic refraction techniques are rarely used today, except for reconnaissance surveys to determine the depth to a high velocity metamorphic or igneous basement below an inadequately known sedimentary basin.

The reflection method, which works on the echo sounding principal (see Figure 2.1), is the most common modern tool for subsurface exploration. The reflection records build up a vertical picture that provides a direct impression of the arrangement of strata, as in a geological cross section. The amount of geological information gained from seismic records has increased greatly over the years by many improvements in electronic processing. It is now possible to identify minor rock formations such as buried sand banks, slump features, and changes of lithology, as well as anticlines and faults. The reflection seismograph has become almost a precision instrument.

In most of the known petroleum-producing basins, almost all of the structural traps of reasonable size have been drilled. Thus, there is currently an increasing effort to design techniques to locate stratigraphic traps. Seismic technology can provide good structural information in more than 90 percent of the sedimentary basins of the world. A few important areas remain in which the best seismic reflection techniques are unable to define potential structural traps, most often because of an irregular sequence of near-surface formations that distort all passing seismic waves. In the rest of the basins, the quality of the structural information ranges from adequate to excellent, with the typical survey defining dozens of subsurface boundaries between different rock types.[1]

In defining stratigraphic traps, lithologic information must be extracted from seismic reflection data. Rocks of different compositions conduct seismic energy at different speeds. As a general rule, velocities are lower in clastic rocks than in carbonates, and the more porous the rock is, the slower the seismic velocity is. A properly programmed computer can sort out combinations of reflection strengths and travel times and produce a list of seismic speeds through the geologic section of a survey area. The speeds bear a relationship to the actual subsurface lithology. This application is in an early stage of development and requires extensive intervention of skilled interpreters. It has been used successfully, but in some areas the quality of the basic seismic reflection data is not good enough to yield a useful listing of sound speeds. In

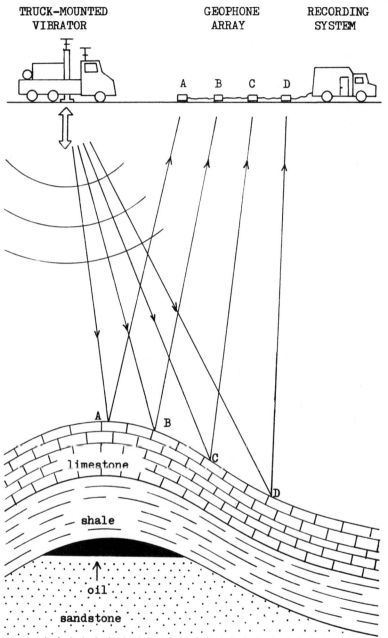

Figure 2.1. Reflection seismic surveying. The length of time required for the sound waves to travel from the vibrator truck to points A, B, C, and D on the anticline at depth and to return to the surface are measured by the geophone array. The anticline is thus identified.

other areas, the results do not always appear to be related to the lithological properties that create traps.

Another related advanced seismic application is known as three-dimensional seismic imaging. In this process, an image of several square miles is achieved and manipulated by computer to produce a cube or a rectangular volume, the interior of which may be studied by dissecting it horizontally or vertically to produce cross sections. In some programs, the plane of cross section need not be vertical or horizontal. Conventional profiles record seismic reflections along a straight line as if all reflections came from directly below the recording line. In reality, seismic reflections bounce back from numerous surfaces on either side of the recording line. Three-dimensional imaging records and manipulates records of these reflections, thus indicating the tilts and dips of subsurface formations. The data for this type of analysis is collected along an intersecting grid in which the geophones are set in a straight line, but the sound impulses are set off along a line perpendicular to them. The midpoint between each shot and each geophone forms a rectangular grid called a crossed array. A series of crossed arrays can be located so that the rectangular grids cover the entire prospect area.

Three-dimensional seismic imaging can be used for field analysis as well as for exploration. A three-dimensional computer representation of the lithology of an oil field can supplement available well information to provide a basis for reserve estimates. As with any new technique, refinements are needed and expected. With improvements in resolution, there can be an increase in the range of geologic problems to which this method may be applied.

In areas where a number of wells have been drilled, the records from these wells can be used for subsurface mapping. Well logs are records of various physical parameters of the strata penetrated by the well, as measured by logging tools run down the hole after the drilling has been completed. By an analysis of these records, it is possible to correlate similar geologic horizons across a study area to determine the geologic structure and stratigraphy. Subsurface maps can be constructed from such information, which may lead to additional structural or stratigraphic drilling prospects.

Techniques to Locate Petroleum Offshore

Geological exploration of the outer continental shelf is accomplished by bottom sampling, shallow core drilling, and deep stratigraphic test drilling. Usually bottom sampling and shallow coring are conducted simultaneously using a small drilling ship. Bottom samples are obtained by dropping a weighted tube to the ocean floor and recovering it by means of an attached wire line. Penetration is usually limited to a few feet, depending upon the local condition of the sea floor. The samples obtained from the tube are useful in identifying the type and origin of the bottom. If the seafloor is composed of sedimentary rock, its geological age can often be determined by the identification of microfossils included in the sample.

Shallow coring is performed by conventional rotary drilling equipment. Penetration is usually limited to the recovery of several feet of consolidated rock. The geological examination of the cores provides useful data regarding the general geology of the area. Deep stratigraphic tests are drilled, utilizing larger offshore rigs, for the acquisition of geological and drilling information and may reach 16,000 feet. Stratigraphic tests are usually drilled off-structure and are specifically not tested for oil and gas. By the use of various well-logging devices and an examination of the drill cuttings and cores, the geological section can be determined. Potential source and reservoir rocks can be studied and parameters such as source rock maturity and reservoir porosity and permeability determined in the area of the test. A general assessment of the potential for petroleum can thus be obtained. In addition, the knowledge of the geological section is an aid to the geophysical studies being conducted at this stage of the offshore exploration process.

Seismic exploration provides additional information at depth by measuring the velocity of shock or seismic waves through the various rock formations beneath the seafloor. The knowledge of the geologic section gained through deep stratigraphic testing is useful in the interpretation of the seismic records. Seismic data from shallow formations is of value in the identification of potentially hazardous drilling conditions, such as surface and near-surface faulting, potential slide areas, or shallow gas pockets. Information of this kind is important in the choice of drilling and production platform location.

For regional and detailed mapping, deep-penetration seismic information is needed. If possible, the data is interpreted by the mapping of at least two seismic reflection horizons corresponding to the depth of expected hydrocarbon production. The seismic records indicate the types of structures (anticlines, faults, salt domes, etc.) that appear to exist in the area and their locations. Such structures become the prime targets for exploratory drilling.

In exploration seismology, energy is transmitted into the earth and the recorded reflections provide information about the subsurface that can be used for the delineation of geologic structures. The most common sources of energy for offshore seismic surveying are air or gas guns, which generate seismic waves without the use of explosives. Guns of various sizes provide enough energy to penetrate as much as 20,000 feet of sediments in most areas. Geophones, which are deployed to detect the reflected seismic energy, are very sensitive instruments enclosed in a cable up to 9,000 feet long. The cable, which is towed behind the survey ship, is a thick, flexible tube containing the geophones and the wires carrying the seismic data to the recorders aboard the ship. It is filled with oil to provide buoyancy and better acoustic coupling with the water and is fitted with stabilizers to control its depth below the surface. The equipment on board the ship records the seismic signals on magnetic tape in digital format. These field data are then processed in a digital computer to eliminate unwanted random energy (noise). After the data have been processed to obtain maximum quality, they are displayed in the form of a vertical cross section that exhibits geologic structure.

Seismic reflections are caused by velocity changes in the rock formations. Increasing velocity difference between two geologic horizons means increasing amplitude of the reflected energy. As seismic velocity in a gas- or oil-saturated sandstone is lower than in either a water-saturated or nonporous sandstone, the presence of oil or gas in a formation will cause a two- to fivefold increase in the amplitude of the reflected energy. By recording and processing such seismic data in a manner that preserves the true amplitudes of the reflections, it is sometimes possible to directly identify oil- or gas-bearing sandstones. These specially processed records, when displayed as cross sections, show strong events when abnormally large contrasts in seismic velocities exist. These strong events are referred to as "bright spots," which is the term used for this method of direct hydrocarbon detection.

A second indicator of hydrocarbons is the polarity of the reflected seismic waves. Seismic waves are transmitted as compressions and expansions of the media through which they pass. The polarity of a wave refers to the direction of its first motion, which depends upon the relative velocities of adjacent media through which waves pass. A low-velocity, gas-bearing horizon can sometimes be distinguished from a dense limestone by the opposite polarization of the reflecting signals.[2]

A final indicator of the possible presence of petroleum is a reflecting interface that is perfectly horizontal. Because geological formations almost always have at least a regional dip, a horizontal interface is evidence of a contact between two fluids, such as gas over formation water or gas over oil. Where rock layers are actually flat, however, the fluid interface cannot be detected.

These direct detection techniques are not applicable in all prospecting situations. They appear to work best in young, relatively uncompacted sediments in offshore basins, such as Tertiary-age strata consisting of sandstones and shales. The techniques are more difficult to apply in the older continental basins, where the geology is often more complex. They are also rather ineffective below 10,000 feet because the acoustic signals become increasingly attenuated in passing through such thickness of rock. The bright spot is essentially a technique for finding gas rather than oil, as the density of gas differs from the density of formation water to a greater degree than does the density of oil. The bright-spot anomaly can generally indicate the presence of petroleum, but not the quantities involved. Bright-spot technology has had many successful applications, but there have also been significant failures. Perhaps the most notable of the failures involved the Destin Dome in the Gulf of Mexico, where the method reportedly indicated relatively shallow gas deposits and was probably responsible for very high bidding on several federal tracts. A succession of dry holes proved that the petroleum potential had been significantly overrated.

In offshore surveying, water depth is measured by bouncing a sound signal off the seafloor and recording the time required for the signal to make a round trip. If two seconds are required the water is about 4,725 feet deep, if one second is required the water is only about 2,360 feet deep, and so forth for fractions of seconds. To obtain information on the thickness of seafloor mud, a stronger echo sounder, sometimes called a mud penetrator, is used. As the survey ship travels, this sounder bounces sound waves off the rocks that commonly underlie the mud of the seafloor (see Figure 2.2). The return signal is graphically recorded on a roll of paper. In this manner a continuous record of the thickness of the bottom mud along the ship's path can be made. This information is of value in the positioning of drill ships and production platforms.

Often a magnetometer is towed behind the survey ship. It detects small warps or anomalies in the earth's magnetic field that are produced by different types of rocks over which the ship passes. These anomalies are indicators of the structure of the formations at depth below the ocean floor.

Survey ships also often contain gravimeters, extremely sensitive instruments that measure slight changes in the force of gravity caused by the ship passing over rocks of varying densities. These anomalies are also indicators of geologic structure.

The geologic, geochemical, and geophysical techniques used to explore for petroleum are very useful as indicators of geological conditions (structures, potential source and reservoir rock, lithology, and so on) conducive to the generation and entrapment of petroleum and in locating potential production prospects. The only direct method, however, of proving a commercial deposit is to drill it and test its fluid content. Without drilling and testing potential traps, the accumulation of

48

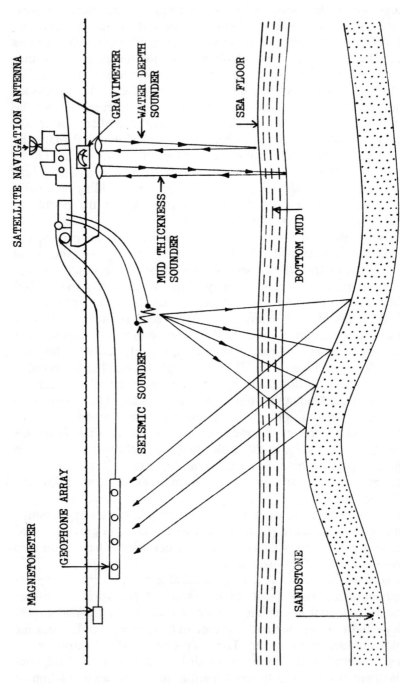

Figure 2.2. Geophysical survey ship showing water depth sounder, mud thickness sounder, seismic sounder, geophone array, gravimeter, and magnetometer.

commercial amounts of petroleum in a basin cannot be determined with certainty.

Drilling Technology

For many centuries holes have been drilled in various parts of the world for the purpose of finding and producing petroleum in its different forms. The first well drilled in the United States for the express purpose of producing oil was the Drake well in Titusville, Pennsylvania. It was completed in August 1859 at a depth of 69.5 feet. This discovery and the publicity given to it resulted in the first American oil boom. During the next few years many other wells were drilled in Pennsylvania and also in New York, West Virginia, and Kentucky in the search for petroleum in commercial quantities. Drilling locations were chosen primarily on the basis of surface oil seeps or where oil had earlier been observed in wells drilled for brines.

The Drake well was drilled by the cable tool method, which is based on a percussion principle in which the rock material at the bottom of the hole is pulverized or broken up by means of a solid steel cylindrical bit with a blunt, chisel-shaped cutting edge. The bit is attached and works vertically at the end of a heavy string of steel tools suspended in the hole by a steel cable. The other end is attached to a walking-beam engine. The reciprocating cutting action is activated by the up and down motion of the walking beam, with the bit chipping the rock by a series of repeated blows. After drilling a short distance, the hole must be emptied of rock chips and fluids. A long, hollow steel cylinder with a valve at the bottom (a bailer) is used for this purpose. The bailer is attached to a wire line and lowered into the hole to recover and remove the rock cuttings, water, and mud that would impede further drilling. Steel pipe or casing is run into the hole when required by hole conditions. The casing is supported at the surface and lowered joint by joint as the hole is deepened. To add a joint, the hole is deepened at least one joint length with an underreaming bit that opens to a larger diameter than the casing, thus allowing the casing to be lowered. A cable tool rig consists of a derrick to support the drilling line, the sand line on which the bailer is run, and the casing line. Although wells have been drilled to as deep as 5,000 feet using this method, cable tool rigs are usually used for much shallower holes.

The hydraulic rotary drilling system (see Figure 2.3) was developed in the United States during the two decades preceding 1900 and was first used to drill water wells. Since about 1927, the development and improvement of rotary equipment and drilling practices have been revolutionary. The petroleum industry now has available a variety of

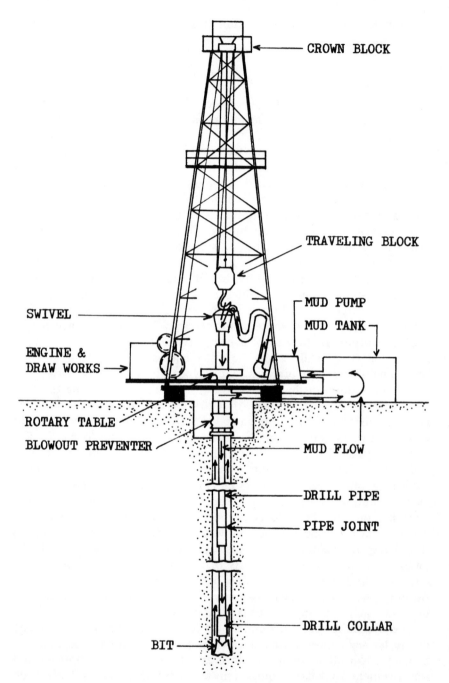

Figure 2.3. Rotary drilling rig. Arrows indicate the direction of mud flow.

rigs designed and built in many sizes and types for use in drilling wells of varying depths in differing environments. Cable tool drilling has declined during this period to a relatively minor role in oil-well drilling.

Rotary drilling, now the most common method of drilling, is a hydraulic process consisting of a rotating hollow column of steel drill pipe at the bottom of which is attached a hard-toothed or diamond-impregnated drill bit. The rotary motion is transmitted through the pipe from a rotary table on the rig floor. As the pipe and the attached bit turn, the bit loosens or grinds a hole in the rock. During drilling a stream of water or drilling mud is in constant circulation down the pipe and out through the bit, from where it and the suspended rock cuttings are forced back up the hole outside the pipe and into storage tanks where the cuttings are removed by a vibrating, fine-wire-mesh screen (the shale shaker) and the fluid is pumped back down the pipe. The heart of the fluid circulating system is the mud pumps. On large rigs, there may be two pumps of 1,500-horsepower size. The hoist (or drawworks) is designed to raise or lower the drill pipe and casing used in a well. On deep wells the casing load can be in excess of 350 tons. The derrick is a fabricated steel tower sufficiently strong to support the loads imposed by drilling operations. Derricks are usually 130 to 145 feet high and are capable of supporting loads up to 500 tons. The drill pipe is rotated by means of a rotary table, located directly above the well and powered by diesel and/or electric motors.

The drill string consists of the kelly, the drill pipe, the drill collars, and the bit. The kelly is the topmost section of the drill string and transmits the drive from the rotary table. It is usually about 40 feet long and square or hexagonal in cross section. It is supported by the traveling block (hanging from the derrick) and slides downward through rollers in the rotary table as drilling proceeds and the hole is deepened. The drill pipe is composed of sections or joints about 30 feet long that are screwed together by special leak-proof couplings. Between the drill pipe and the bit are drill collars. These are heavy-walled cylinders of steel, also about 30 feet long, which concentrate the weight immediately above the bit and maintain the drill pipe in tension. (Otherwise the drill pipe will buckle in compression, causing hole damage, excessive wear, and fatigue failures.) Both drill pipe and drill collar connections are precision-made to provide pressure-tight seals at the connections and rotational stability and strength when in the hole. Great care must be exercised when handling them to prevent thread damage, which would destroy the pressure seal and cause rotational failure in the pipe.

At the bottom of the drill string is the bit. Some drag or fish-tail bits are still used in very soft formations (in the Gulf Coast, for example), but the majority of bits currently in use are of the tri-cone type. The tri-cone bit utilizes three forged-steel conical cutters whose axis of rotation is slightly offset from a horizontal plane. As the bit is rotated, the cutters also rotate and the cutter teeth dig and gouge out the rock. Long-tooth bits are used for soft formations and bits with shorter teeth for harder formations. In hard rock the teeth of the bit chisel the rock and heavy weight is required for drilling. There are two types of tri-cone bits, the conventional and the jet. The conventional tri-cone has fluid passages through the center directed on the cutters. The jet tri-cone has variable-sized jets that direct the fluid circulation to the bottom of the hole to jet away the cuttings. In soft formations this action erodes the bottom of the hole, increasing the penetration rate and prolonging the life of the bit. Bits with diamonds set into the face are also used for drilling hard formations. Though much more expensive, they drill farther before having to be replaced, thus saving rig time required to change bits.[3]

Drilling muds, fabricated to lubricate the drilling bit and drill string and to remove cuttings, are necessary also to prevent caving of formations and to control subsurface pressures. To assure pressure control, however, blowout preventers are used. They are connected to the top of the wellhead during drilling and can be shut in to control the well if gas or oil is encountered at pressures higher than those exerted by the column of drilling mud in the hole. The blowout preventer is attached to the surface casing, a relatively large-diameter, thin-walled steel pipe set and cemented into the hole to act as a liner to prevent caving and contamination of aquifers and to control well pressures. Often several strings are run, each longer than the one run before. In deep wells, as many as five strings are sometimes needed, depending on the formation conditions encountered.

Blowout preventers usually consist of two or three ram types and one bag type, with working pressures between 3,000 and 10,000 pounds per square inch, depending upon well depth and expected pressures. The topmost preventer is usually the bag type, consisting of a rubber ring that can be inflated with fluid by a piston and squeezed around the drill string, providing a seal between the string and the casing. By adjusting the pressure on the piston, the drill string can be moved through the rubber ring, which adjusts the ring's shape as required.

The lower two pipe rams, which can be closed around the drill pipe, are more secure. These are hydraulically operated with one ram on each side of the bore hole. The leading edge of the rams has a recessed face and rubber seals to fit around the body of the pipe. If these alone are activated, it is still possible for the well to blow out through the drill pipe, if the kelly is not attached. Under this circumstance a third ram, the blind ram, may be activated, which deforms or crimps the pipe, closing the hole completely. An alternative form, the shear ram, shears off the drill pipe, allowing it to drop into the hole. As it is difficult to reestablish control of a well after the drill string has been lost, the decision to use the shear ram is not made lightly. Recently an internal preventer has been developed to close off the space inside the drill pipe. The blowout preventers are hydraulically operated from a control panel on the rig floor and a second remote control panel is situated some distance from the rig for use in emergencies when the rig is unapproachable.[4] Connected below the blind ram are the kill and choke lines, which are used to control the well in the event of an imbalance between formation and drilling fluid column pressure. Both lines are high-pressure two- or three-inch pipe. The kill line is connected to the mud circulating system and the high-pressure cement pumps. The choke line leads to a back-pressure control unit and mud degasser.

An impressive development of the drilling industry has been the increase in the depth of wells. Through 1947, only 14 domestic wells had been drilled below 15,000 feet. A few years later a 20,000-foot drilling depth was reached. Today, wells drilled between 15,000 and 20,00 feet are not uncommon, numbering over 600 per year. The first domestic well to reach 30,000 feet was drilled in 1972, and the deepest well in the United States, over 31,000 feet deep, was drilled in 1974. Both of these wells were drilled in the Anadarko basin of Oklahoma, and are the only domestic wells to date that exceed 30,000 feet in depth. The deepest well in the world is thought to be in the Soviet Union on the Kola peninsula. It has been drilling since 1970 and recently reached a reported depth of 34,448 feet.[5] Giant rigs with 152-foot derricks capable of supporting loads up to 1,250 tons are required to drill to 30,000 feet. The casing load capacity of such rigs is 1,000 tons. Many such rigs are now drilling. Their ultimate drilling depth is not limited by drill-string weight, but rather by the weight of the casing up the hole.[6]

In the United States there are over 65 rigs rated to at least 30,000 feet. The capability to drill below 30,000 does not mean that each well drilled by these rigs will be in the 30,000-foot range. Most wells drilled by such rigs are between 20,000 and 30,000 feet deep. In fact, depending upon local conditions, a 15,000-foot hole may be much more complicated and costly, even on a trouble-free drilling basis, than a 25,000-foot hole in a different area. Many deep wells are abandoned short of their original objectives because of various drilling-related problems such as lost mud circulation, formation instability, unanticipated high and/or low formation pressures, mechanical difficulties, stuck pipe, etc. The results of failing to reach the drilling objective are very costly, and the prospect remains unevaluated.[7] In spite of such problems, the rotary

Figure 2.4. Rotary rig drilling in the Arctic. Photo courtesy U.S. Geological Survey.

rig is a very reliable type of drilling equipment. The rotary drilling method has proved so effective that the vast majority of drilling is done with rotary rigs. The United States has over 4,000 active rotary rigs engaged in domestic drilling (see Figure 2.4).

Fundamental studies on the principles of rock breakage and the effect of drilling fluid on rock-cutting removal have contributed to the advance of rotary drilling. Much research, however, has been directed at finding better methods of drilling holes. Finding a way to drill without having to rotate a long, heavy drilling column and then remove it when its cutting tool is worn has been an objective of such research. Avoiding rotation has been partially resolved by the advent of the turbodrill, pioneered in the Soviet Union and now popular in Europe. The turbodrill is a mud motor. Drilling mud pumped down the nonrotating drill pipe turns a turbine or a screw-type drill at the end of the drill string, which is attached to a bit. For continuous drilling, turbodrills may be 40 feet long and contain up to 150 stages. Rotation is rapid and best suited for a diamond bit in hard formations. Normally, a turbodrill is not used with a tri-cone bit because of the rapid wear caused by the high rotational speed.

In the Soviet Union, experiments are also being carried out with a down-hole electric motor driving the bit. Other research by the international petroleum industry concerns various methods of rock attack including mechanical, thermal, and electrical abrasion. Tests have been conducted with high-pressure bits operating at 10,000 to 15,000 pounds per square inch, which drilled rock formations much faster than conventional bits. Other research is aimed at reducing the trip time necessary to pull the long drill string from the hole and then return it, either to change bits or to conduct the various tests necessary for the location of oil-bearing formations. Reduced drilling time, accomplished by improved drilling methods and faster trip times, will result in lower drilling costs.

During the drilling of a well, it is often necessary to obtain a core of the rock penetrated, particularly of the oil reservoir. A core is an undisturbed sample of the formation that can be used for identification and dating of the rock and for measurements of permeability and porosity and oil and water saturation. Cores are usually taken on exploration wells and on early-development wells. They are cut with a special core barrel in multiples of 30 feet (to as much as 90 feet, depending upon the formation). As successive casing depths are reached, electric, sonic, and radioactivity logs are recorded before the casing is run and cemented. The logging tools are run into the hole on a multicore cable from the winch of a special self-contained logging unit. The logging tools are run to the bottom of the hole and the logs are recorded as they are pulled out. The signals from the logging tools are recorded photographically on a continuous film strip in the form of graphs. The electric logs measure the resistivity of the formations and their spontaneous electrical potential, the sonic logs record the transit time of a sonic impulse through the formation adjacent to the tool, and the radioactivity logs measure the natural radioactivity of the formations and also the radioactivity induced by a radioactive source in the tool. From this information it is possible to measure porosity, fluid resistivity, and fluid saturation of a formation and to determine lithologic changes for purposes of geological correlation. Other logs sometimes run are the cement bond log, which measures the bonding between the casing and cement and the cement and the formation, and the dipmeter, which records formation dips. Continuous hole directional and deviation surveys can also be run.[8]

The drilling process requires skill and experience, although almost all rigs have a weight indicator, which gives the weight on the bit, and a pump pressure gauge. A well-equipped modern rig will also have a pump-stroke indicator, a rotary-speed and torque indicator, and a mud-fluid–level indicator to help the driller. New systems are being developed that use a microcomputer to collect additional drilling data and display the information on a cathode ray tube, which allows the driller to set alarm limits on any monitored parameter.

If it appears that the hole has encountered a potential petroleum reservoir, a drill-stem test is a temporary completion of the well during drilling to measure formation pressure and obtain samples of the formation fluids for evaluation. A test string of pipe is run into the hole and packers are set to isolate the test zone. After pressure and flow testing, a sample of reservoir fluid is trapped in the pipe. This information is of value in determining whether the reservoir warrants development as a commercial petroleum field.

Offshore Drilling Technology

In 1980 about 23 percent of the world's oil was produced offshore. There was an increase of about 9 percent over 1979 offshore production, while onshore production fell. Drilling methods and much of the equipment used offshore is similar to that used onshore, including the rotary drilling method. Offshore, the mud is circulated down the pipe and out the bit during drilling and is returned back to the surface by means of a marine riser. Marine risers were developed to conduct the drill string from floating rigs to the hole being drilled and also to conduct the drilling mud back up to the rig. They are designed to permit some lateral and vertical movements during drilling operations without breaking off the drill string. In addition to removing the cuttings, the drilling mud helps prevent blowouts by counterbalancing formation pressures. The necessary balance is accomplished by regulating the weight of the mud and the mud-flow rate. The other safeguards to assist in the prevention of blowouts are casing and blowout preventers. On offshore bottom-standing platforms, the blowout preventer stack is attached to the top of the surface casing just beneath the rotary table on the rig floor. In the case of floating rigs, the stack is attached to the top of the surface casing on the seafloor and is hydraulically activated and controlled from the rig. Blowout preventers are activated manually, not automatically.

Unlike onshore technology, offshore drilling operations require a platform to support the drill rig and its associated equipment. There are four basic types of offshore exploratory drilling platforms: barges, drill ships, jack-ups, and semisubmersibles.

Barges were used extensively in drilling the first wells in shallow waters in the Gulf of Mexico. Compartments in these early barges were control-flooded to set the barge on the seabed. This method of exploration drilling with submersible barges was limited to water shallow enough

Figure 2.5. Drill ship. Photo courtesy U.S. Geological Survey.

(less than 75 feet) to permit the upper structure of the barge to rise above the water to a height that would permit drilling. In contrast, most new barges are floating. Although not shaped like a ship or self-propelled, they are used much like a drill ship and are capable of drilling in water depths of 600 feet or more. The depth limitation is dependent upon the anchor and chain system used for maintaining position. Barges also are limited by weather conditions, due to their shape and consequent poor motion characteristics.[9]

Drill ships (see Figure 2.5) have the same general lines as traditional merchant ships, are self-propelled, and, therefore, can move from one drilling location to another without assistance. Positioning is accomplished by a dynamic positioning system (a series of propellers or thrusters coupled to sensors that detect and compensate for movement) or a mooring system of chains and anchors (see Figure 2.6).

In 1979, a drill ship established a world water-depth record for conventional petroleum exploration by drilling a 20,023 foot hole in 4,876 feet of water off Newfoundland. The hole was dry and therefore abandoned, but the project was considered a significant success. The ship is equipped for drilling in 5,000 feet of water and had set prior deep-water drilling records in 4,346 feet of water off the Congo and in 4,441 feet off Spain. The ship is dynamically positioned with six 3,000-horsepower variable pitch right-angle thrusters and two 8,000-horsepower main screws. Positioning is controlled by a computerized automatic station-keeping system using a dual acoustic reference with riser angle indicators and backup sensors. The drawworks is rated at 3,000 horse-power and the 170-foot derrick at 665 tons.[10]

There are 22 drilling barges in operation in the world, more than half of which are in Venezuela. There are 62 drill ships currently in operation, with more in Southeast Asia and the Far East than in any other region. Twenty-one of 22 existing submersible barges are in the United States.

Jack-up rigs (see Figure 2.7) are drilling platforms with legs that can be moved up and down. When the legs are extended, the platform can elevate itself above the ocean surface and temporarily become a bottom-standing platform. By retracting its legs, the jack-up becomes a floater and can be moved from one location to another with assistance. Jack-ups can drill in water depths up to about 350 feet. They are the most popular of the offshore drilling units, with 291 operating throughout the world. Of these, 115 are in U.S. waters, mostly in the Gulf of Mexico. An advantage of jack-up rigs is that when they are jacked up within their water-depth range, the effect of heavy seas is negligible. Also, their steel and engine requirements (and thus costs) are considerably less than the corresponding requirements of drill ships and semisub-mersibles. The main disadvantages of jack-up rigs are the difficulty of moving onto location and the fact that many of the units cannot jack up when long-period swells are running. Another problem is an awkward configuration during transit: the heavy steel legs, when projected upward, can cause a degree of instability. These problems, however, are not insurmountable, and jack-ups have good operating records. Recent designs make jack-ups operational in progressively harsher environments. One new design, with legs 452 feet long, is for 300-foot water and will be able to withstand the combined forces of 85-foot waves and 90-knot winds, along with a 1-knot current.

The semisubmersible (see Figures 2.8, 2.9, 2.10, and 2.11) is the newest type of offshore drilling rig and is the best suited for severe

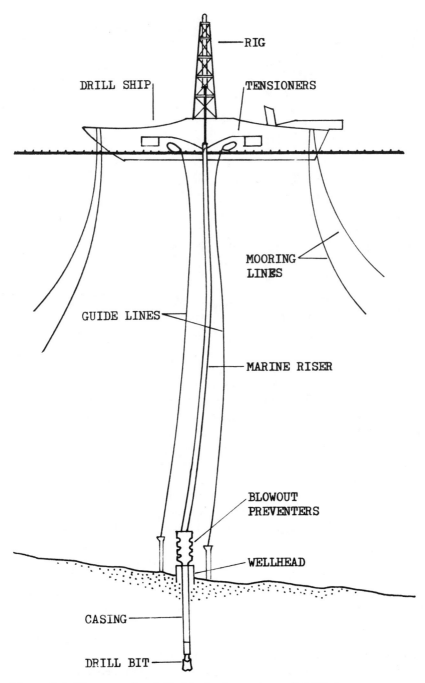

Figure 2.6. Diagram of a drill ship in the process of drilling.

Figure 2.7. Jack-up rig. Photo courtesy U.S. Geological Survey.

weather conditions. These rigs have a platform deck supported by columns connected to large underwater displacement hulls or large vertical caissons, or a combination of both. The columns, displacement hulls, or caissons are flooded on site to reduce the force of the waves by locating the major buoyancy members below the sea surface or below the level of wave action. The units are considered virtually transparent to waves of normal period as the water plane area of the columns is usually less than a third that of a comparable drill ship. The flooded members require rigid maintenance, however, as the failure of the remote control valves on a ballast tank could cause the rig to turn over.

Semisubmersibles may or may not be self-propelled. Most are positioned with mooring systems, but some are equipped with dynamic positioning systems (thrusters and sensors). There are 106 semisubmersibles drilling around the world, with most operating in the North Sea. The large semisubmersibles were designed primarily for rough waters and difficult environments such as the North Sea and offshore Alaska. The primary disadvantage of the semisubmersible is its high cost. Several large semisubmersibles have been designed for operation in 3,000 to 4,000 feet of water.

Figure 2.8. Semisubmersible drilling rig. Photo courtesy U.S. Geological Survey.

With increased interest in offshore drilling, many new drilling units are on order. In early 1981 there were 143 jack-ups, 43 semisubmersibles, 8 submersibles, and 4 drill ships—the most units ever—under construction. The construction time for a jack-up is a little over a year and for semisubmersibles about two years. With the large number of jack-ups that will soon be available, it is possible that there will be a surplus in shallow-water rigs in the next year or two.[11]

As drilling moves into ever deeper waters, the main technological problem appears to be riser technology, especially for water depths in excess of 10,000 feet. The problems associated with a 10,000-foot riser are: increased trip time to deploy and retrieve the riser, making the operation vulnerable to sudden storms; a resonance period close to that

Figure 2.9. Semisubmersible drilling rig. Photo courtesy U.S. Geological Survey.

of waves, decreasing its stability; and danger of unexpected disconnection while in tension, leading to extensive shipboard damage.[12]

Another problem as exploration and production move to deeper and stormier portions of the sea is safety. In the twenty-six years since offshore drilling has become common from floating rigs, 47 of these rigs have been lost at sea. In the past ten years, offshore accidents have claimed 331 lives.[13] Eighty-four workers died when a semisubmersible capsized and sank in a storm off Newfoundland in February 1982. This tragedy was exceeded only by the loss of 123 men in 1980 when a floating dormitory for oil workers collapsed and sank in the North Sea off Norway. It is evident that safety procedures and perhaps even design factors will have to be improved to cope with the more difficult offshore drilling environments.

Figure 2.10. Derrick on a semisubmersible drilling rig. Photo courtesy U.S. Geological Survey.

Exploration in Arctic waters creates additional problems such as icebergs and superstructure icing of the drill ship. The riser is usually disconnected when an iceberg is between one-half mile and one mile away from the drill ship. If the draft of the iceberg is such that it could hit the blowout preventer on the seafloor, the disconnect of the riser must be made below the blowout preventer stack and a cement plug and appropriate retainers must be used before disconnecting. To avoid this extreme measure, an attempt to tow the iceberg away from the rig may be made. If the iceberg is smaller than 1 million tons, a standby supply vessel can usually do this successfully. For icebergs above 2 million tons, towing will not be effective. Dynamically positioned drill ships are often used in iceberg-prone regions because they permit quick

Figure 2.11. Drill pipe in derrick on semisubmersible drilling rig. Photo courtesy U.S. Geological Survey.

disconnection and reconnection and eliminate the loss of time and other problems associated with anchoring. Superstructure icing greatly affects a vessel's seaworthiness by raising the center of gravity, which results in a loss of stability.

In near-shore Arctic drilling, artificial islands are sometimes used instead of drill ships. They are constructed with granular fill that is either hydraulically or mechanically placed. Additional offshore platforms have been constructed of ice (see Figure 2.12). A stable pan of multiyear ice at least one mile in diameter and averaging five feet in thickness is required over the drilling location before ice-platform construction can begin. Construction begins in November (in the Canadian Arctic Islands) with the excavation of a hole cut through

Figure 2.12. Arctic ice drilling island. Photo courtesy U.S. Geological Survey.

the natural ice to the water and cribbed with planks. This hole will be used, subsequently, as the drilling pool. The snow is then removed from the ice in a 400-foot diameter area, which is then flooded with water. Flooding and freezing are accomplished in successive one- to two-inch layers, with an average ice buildup of about three inches per day until the platform is about twenty-two feet thick. The main disadvantage of such ice platforms is that, in the Arctic Islands where they are used, they can be utilized only during the five-month period from January to May when the natural ice is not moving and the artificial ice platform has had time to be thickened sufficiently to carry the weight of the drilling rig. These factors place a limit on the size of the rig and, thus, on the depth that the well can be drilled. Also, there may be areas farther from shore where horizontal ice movement is too great for the system to be used at all.[14]

Production Technology

Primary Recovery

The proper completion of an oil well is essential for the control of, and maximum recovery from, a reservoir. An inherent disadvantage of the rotary drilling system is that the pressure of the drilling fluid must be greater than the pressure of the formation fluids when drilling through petroleum-producing formations, or the well will be out of control. The water-base drilling muds penetrate the reservoir rock and often clay particles and mud-weighting materials are carried into the formation, reducing permeability. This is termed "water blocking" or "mudding off" and may be permanently detrimental to oil production. To overcome the effects of such infiltration, cable tools are sometimes used to drill into reservoirs, especially when formation pressures are low. Another method of protecting the reservoir is the utilization of oil-base drilling fluids when drilling oil formations that are subject to damage by water infiltration. In certain cases, surface-tension–reducing agents or sulfonates are used to remove drilling muds from the reservoir.

The casing program is of great importance in the proper completion of a rotary-drilled well. There are two casing programs in use in completing wells drilled by the rotary method when oil production is from a single horizon. In both of these programs it is necessary to set and cement surface casing to protect the fresh water sands from contamination and to maintain pressure control. In one casing program, it is the practice to set and cement a string of casing immediately above the oil-producing horizon and then set a perforated liner, if needed, or produce from the open hole. The other casing program is to drill and set and cement the casing through the producing zone, and then to perforate the casing opposite the horizon marked for production. A perforated liner can then be set if needed. The most common casing perforator was once the single-, double-, or four-knife casing splitter, which cuts slits of any desired length. The in-the-hole gun perforator was developed later and consists of an alloy steel cylinder that contains chambers with powder charges and bullets. The gun is lowered into the hole opposite the producing horizon and the shots can be fired individually or as a group by electrical controls at the surface. A modification of this method has evolved using shaped charges that need no bullets to perforate the casing. The perforations are made by focusing high-velocity force streams against the casing. This method has improved the depth of penetration into the reservoir. If petroleum is to be produced from more than one horizon, the casing is perforated opposite each productive zone. Sometimes multiple completions are accomplished in a well without mixing the produced fluids. This involves setting separate strings of tubing, each to the level of its production horizon, separating them with packers, and perforating them.

Typically, oil wells are removed from production long before all of the in-place oil has been recovered from the reservoir. The fraction of in-place oil that will flow unaided or can be pumped from the reservoir rock matrix to the surface varies widely, from as high as 80 percent to as low as 5 percent. The factors that determine the efficiency of such primary recovery are the energy (pressure) contained in the reservoir, the viscosity (resistance to flow) of the oil, and the permeability (ease of flow permitted) of the reservoir rock.[15]

In general, primary recovery averages around 20 percent of the in-place oil. In the United States, however, average primary recovery has been somewhat higher, about 28 percent. This is due to the large, highly permeable reservoirs with strong natural water drives discovered in East Texas and southern Louisiana and to the super-giant Prudhoe Bay field, with its moderate water drive. Such reservoirs allow a more efficient primary production than others with less porosity, permeability, and reservoir energy, which are common in some oil-producing regions.[16]

Due to certain physical conditions, a part of the original gas volume in a reservoir is not recoverable. These conditions include the heterogeneity of the reservoir and the trapping of some gas in the reservoir by water influx. In addition, when reservoir pressures become depleted a point is reached at which the gas can no longer be economically produced. In spite of these conditions, the percentage of in-place gas recoverable during the economic life of a reservoir is quite high, several times higher than the recoverable percentage of crude oil. Depending upon the permeability of the reservoir, recovery can be as high as 75 to 80 percent of the original in-place gas.

In 1860 a shot of black powder was exploded alongside a string of tools stuck in a water well to loosen them, and there was a noticeable increase in water production following the shot. The results obtained were responsible for experiments with shooting oil wells between 1860 and 1865. In 1865 a well near Titusville, Pennsylvania, was experimentally shot with two gunpowder torpedoes. After the explosions, the well began to flow oil and paraffin. The next year a dry hole was shot with a torpedo and production was initiated. A second shot a month later resulted in a recovery of 80 barrels of oil per day. This established the torpedo beyond question, and there was an immediate demand for shooting throughout the Titusville oil-producing area.

Later, liquid nitroglycerin replaced gunpowder and in 1926, during the early development of the Permian basin in West Texas, it was the general practice to use liquid nitroglycerin for shooting oil wells. Thereafter, there was a gradual change to the use of solidified nitroglycerin. The formation to be shot had to be directly exposed to the explosion and any formation, other than an unconsolidated one, was considered a prospect for shooting to increase production. The shooting produced an enlargement of the hole and many cavings that had to be recovered. As time went on, safer methods of detonating the shots were developed, as premature explosions were a hazard when using nitroglycerin. Today, shooting has largely been replaced by more modern recovery methods.

Although acid treatment of oil wells was attempted as early as 1895, the process was infrequently used for the next thirty years. Then, in 1932, several wells in limestone reservoirs in Michigan were successfully stimulated with hydrochloric acid treatments. These successes increased interest in acidizing to improve well productivity and several companies were organized to provide the service commercially. Since its first commercial use in 1932, hydrochloric acid has remained the primary acid treating agent for oil wells. It is effective and has a relatively low cost.

Although the basic chemistry of acid treatment has been established for many years, the economics and effectiveness of each individual acid treatment are dependent upon the local condition in the bore hole and in the formation. During the early years of acidizing it was assumed that acid uniformly entered and enlarged the formation pores and thus increased the flow of the in-place oil. It has recently been demonstrated, however, that the acid is expended very rapidly in a matrix of carbonate rock and penetrates very little beyond the well bore. Thus, its primary effect is to remove well-bore damage.

In tight carbonate rock, it is difficult to avoid pressure parting and therefore acidizing usually occurs in natural or hydraulically created fractures. In most cases deep penetration of acid through fractures is desirable to attain large productivity increases. Since the penetration of spent acid into the formation provides little benefit, it has been found that an increase in the injection rate provides deeper penetration before the acid is spent. Fluoride chemicals are sometimes added to the acids to speed up the rate of reaction on dolomites and siliceous rocks. To this fortified solvent, various surface-active agents, demulsifying compounds, and corrosion inhibitors have been added to produce a fluid that is used to dissolve the silicates present in oil-bearing formations and drilling muds. This well-treating mixture is known as "mud-acid."

Effective acid treatment often requires proper conditioning of the reservoir rock before and after acidizing. Many field treatments have been successful only because of the sandwiching of the acid between chemical rock-conditioning agents. These fluid conditioners, such as mud-acids, may be solvents, chemical preflushes, or even hydrochloric acid preflushes. One surfactant, mutual solvent ethylene glycol mono-butyl ether, demonstrates a wide range of improved stimulation results and has become commonly used in field treatments.

The hydraulic fracturing technique for stimulating the production of oil wells is one of the major developments in petroleum engineering. Before 1950, acidizing was the primary method used to stimulate well productivity. However, acid stimulation of nonreactive reservoirs, such as sandstones, was generally ineffective. Since hydraulic fracturing was introduced, it has been successfully employed in both acid-insoluble and acid-reactive reservoirs.

The occurrence of pressure parting of formations during acidizing, waterflooding, squeeze cementing, and drilling operations had long been recognized. During all these operations it had commonly been observed that below a certain injection pressure, the formation would accept only nominal amounts of fluid; but with a small increase in pressure, increasing amounts of fluid could be injected with little change in pressure. In 1950, the concept of pumping a viscous liquid down a bore hole at a high rate to build up pressure to rupture the reservoir rock was introduced. Sand was added to the liquid to prop open the created fractures. In effect, this process widened the well bore. The viscous fracturing fluid was designed to break down to a thin liquid to facilitate effective cleanout of the fractures and the well bore. Even in the initial tests, this new process was successful in producing significant recovery increases in one-half of the wells tested.

Fracturing technology developed rapidly. Improvements in pumping equipment and the introduction of friction-reducing additives led to increases in pumping rates from less than 5 barrels per minute to as high as 400 barrels per minute. The evolution of fracturing fluids contributed significantly to the rapid development of hydraulic fracturing and to the improvement of stimulation results in a variety of formations. The first treating fluids were napalm gels, but their limitations were soon recognized and lease oil or water was substituted. An important development in the trend toward higher pumping rates was the introduction of fluid-loss control agents along with the friction-loss reducers. In recent years, however, fundamental studies on the importance of fluid properties on fracture width and extent have reversed the trend toward high injection rates and have led to the use of very viscous fluids (either oil base or water base) pumped at low rates. These viscous fluids create wide fractures, effectively carry high sand volumes, and improve fluid-loss control. Some of the fluids commonly used at present are guar gels, cross-linked guar gums, viscous oil-external emulsions, cellulose polymers, and gelled oils.

After the use of hydraulic fracturing became widespread, the most prevalent description of fracture orientation was that pressure parted a formation along a bedding plane and lifted the overlying rocks, resulting in a horizontal "pancake type" fracture. Later it was realized that breakdown pressures were often significantly less than predicted on the basis of overburden weight. Studies revealed that when pressure is applied to a bore hole, the fractures created should be approximately perpendicular to the axis of least stress, regardless of the type of fluid used. Thus, in tectonically relaxed areas of normal faulting, the hydraulic fractures are vertical and are formed at injection pressures less than the total overburden pressure. In tectonically compressed areas, the hydraulic fractures are horizontal and formed at pressures equal to or in excess of the total overburden pressure. Most hydraulic fractures in deep wells are vertical, but horizontal fractures are most common in shallow wells.[17]

Gas wells often flow and do not require artificial lifting. If the gas is not associated with oil it expands out of the reservoir in response to the drop in pressure in the well bore, which can be enhanced by compression. As the gas is produced, the reservoir pressure declines. If the gas reservoir system has a natural water drive from below, reservoir pressure is maintained during gas production by the movement of water into the vacated pores. Thus, the wells can produce at fairly steady rates until depletion. Care must be taken, however, that production rates are not so high as to cause damage to the reservoir by water infiltration.

Oil wells sometimes flow (because of dissolved gas drive, gas-cap drive, or water drive from below) (see Figure 2.13) but usually require an artificial lifting system. The system chosen depends upon the depth of the well, the nature of the reservoir, the gas/oil ratio, the viscosity of the oil, and the costs involved. The first "gusher" in the United States was Spindletop in Texas. It was drilled in 1901 and was estimated to have flowed initially at 75,000 barrels per day. The early wells in Pennsylvania did not flow and, by 1869, the rod-pump system was widely used. The cable tool drilling rig was left on site and the walking beam ran the pump. The walking beam produced the reciprocating movement that moved the rods up and down and activated the pumping action. Currently most of the oil lifting is done by some variety of the "horse head" walking-beam system (originally powered by steam and more recently by an internal combustion engine or an electric motor). The actual job of lifting the oil to the surface is done by the sucker

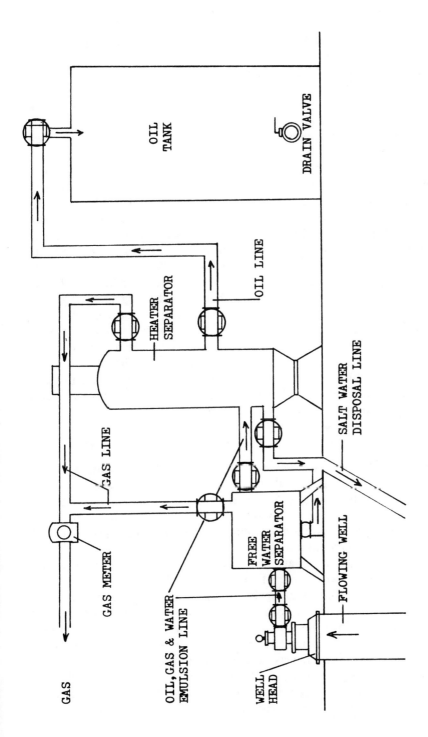

GAS

GAS METER

GAS LINE

OIL, GAS & WATER
EMULSION LINE

WELL
HEAD

FREE
WATER
SEPARATOR

FLOWING WELL

HEATER
SEPARATOR

OIL LINE

SALT WATER
DISPOSAL LINE

OIL
TANK

DRAIN VALVE

Figure 2.13. Flowing well producing oil, gas. and water showing separators, oil tank, gas line, and water disposal.

rods, which act as flexible springs to move the oil up through a series of valves and ball and seating cups.

Another method of bringing oil to the surface is an electrically powered centrifugal submersible pump. The pressure created by the rotation of the pump's impellers forces the fluid to the surface. Submersible pumps are quite efficient at shallow depths (250 to 25,000 barrels per day), but their maximum volume drops rapidly as well depth increases.

Subsurface hydraulic pumping uses the oil in the reservoir to force oil to the surface. As with the submersible pump, this system utilizes a pump at the bottom of the well. The pump is driven by a hydraulic motor. On the surface, a standard engine pump draws crude oil from the top of a settling tank and forces it down through the tubing to drive the hydraulic motor, which in turn drives the down-hole pump. The driving oil is exhausted from the motor into the well bore where it mixes with the reservoir oil, and the pressure from the down-hole pump lifts both to the surface through a second string of tubing. One barrel of power oil will usually lift one barrel of reservoir oil. With this system, high-fluid volumes can be produced from great depth.[18]

In fields where a large supply of associated gas is available, a gas lift may be used. Gas is introduced into the well in the existing space between the tubing and the casing. A series of gas-lift valves, in place along the tubing, close as the gas enters the next lowest valve. This forces the enclosed oil up to the surface.

Secondary Recovery

Primary recovery includes such bore-hole techniques as acidizing, fracturing, and pumping and in the United States averages about 28 percent of the in-place oil. Secondary recovery methods can be used to increase the percentage of oil recovered from a given reservoir. These methods attempt to maintain or restore reservoir pressure by the injection of gas or water (waterflooding). Depending upon reservoir conditions and oil properties, secondary recovery methods can improve recovery to between 30 to 50 percent of the in-place oil.

Waterflooding of oil reservoirs became widespread when it was realized that oil production was so prolific in many of the giant reservoirs of East Texas because of the natural influx of water from adjacent aquifers. Waterflooding could partially duplicate this natural water drive artifically in reservoirs in which it was absent. The primary recovery efficiency in the East Texas clay-free sand reservoirs may exceed 70 percent, but similarly prolific reservoirs occur widely only in southern Louisiana. In the United States as a whole, waterflooding raises the recovery efficiency by a factor of 1.5 to 2.0 over natural solution gas drive to an average recovery of about 40 percent of the oil in place.[19]

Some reservoirs, because of unfavorable physical characteristics of either the rock or the contained oil, do not respond to waterflooding. The production history of a reservoir that has been successfully waterflooded usually shows a primary peak, followed by a lower secondary peak relatively soon after the onset of waterflooding. Shortly thereafter, the production of water increases rapidly at the expense of oil production. The continued economic life of the field depends upon the income from the declining oil production compared to the costs of treating the injected water to reduce corrosion and prevent the growth of bacteria and of the lifting and disposal of the produced brine. The domestic oil industry produces about 6 to 7 barrels of water for each barrel of oil.

Waterflooding is dominant among the fluid injection methods, and its widespread use is accounted for by the availability of water, the relative ease with which water is injected (because of its hydraulic head in the injection well), the ability with which water spreads through an oil-bearing formation, and water's efficiency in displacing oil. Many of the early waterfloods occurred accidentally through leaks from shallow water aquifers or by surface water entering drill holes. In the earliest method of intentional waterflooding, water was injected into a single well and, as the water-invaded zone increased, the adjacent watered-out wells were used as injection wells to further extend the area of water invasion. This is known as "circle flooding." Later, water was injected into a reservoir in every second well (known as a "five-spot" pattern) (see Figure 2.14) or in rows of injection wells separated by rows of production wells. Recently, however, it has become common to inject water around the periphery of an oil field so that it displaces the oil upward in a manner similar to a natural water drive. This is sometimes coupled with the injection of gas into a gas cap at the top of the reservoir to aid in pressure maintenance.

In a waterflood, a region of increased saturation (bank of oil) is pushed ahead of the injected water to the producing wells. With continued production, the percentage of water produced (water cut) gradually increases until the well becomes uneconomical to operate; production is then ended. Maximum efficiency in oil recovery is achieved if the waterflood front remains stable and if the injected water, because of its greater mobility, does not bypass parts of the oil bank in small streams (fingers). Waterflooding averages only about 40 percent recovery of in-place oil because the pores within the reservoir are not of uniform size and the water is usually less viscous than the oil. Since the fluids move faster through the larger pores, some of the more mobile water

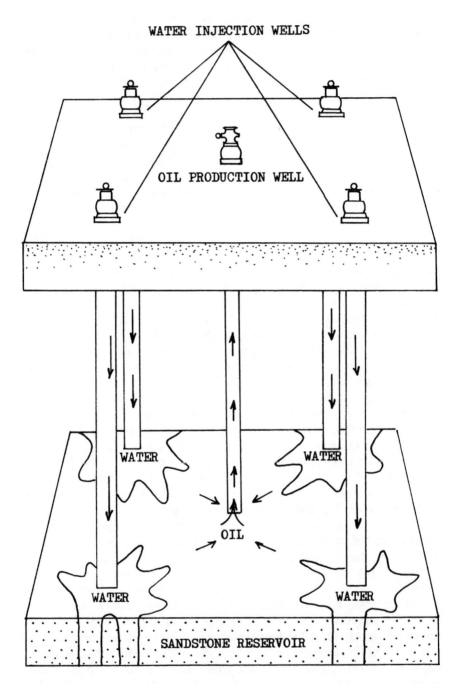

Figure 2.14. Five-spot pattern waterflood.

tends to flow as a finger through the oil, thus bypassing it. This effect is somewhat mitigated because water rather than oil preferentially wets the minerals of reservoir rock and as a result water and not oil is taken up by the finer pores. As the oil is forced through pores already filled with water, oil globules can be cut off from the oil stream. Once an oil globule is separated from the stream and surrounded by water in a pore, the pressure required to deform it so that it can squeeze out of the pore cannot be attained without rupturing the reservoir itself.[20] The main efforts in recent waterflooding practice have been aimed at improving the volumetric sweep efficiency by properly locating the injection wells and by controlling the injection rates.

Gas injection is the oldest of the fluid injection processes. It was used because operators originally believed that injecting water into an oil reservoir might result in a loss of production. Although it is now recognized that waterflooding results in better recovery efficiency because of its lower mobility ratio, gas injection may continue to be used under certain conditions. Fields with low oil saturation gas caps are candidates for improved recovery by crestal gas injection into the gas cap. Gas injection maintains a relatively high pressure gradient on the oil phase and the oil is produced faster and in greater quantity. Gas, however, is becoming more valuable, and its use for injection would have to be weighed against its immediate sale value.

Offshore Production Technology

If commercial accumulations of oil or gas are discovered and defined during the exploratory phase of offshore operations, the development phase begins. Actually, exploratory and developmental activities overlap. While exploratory wells are being drilled to determine the extent of the field and its recoverable reserves, some of the early exploratory wells may be in the process of being completed as production wells as the production platforms are put into place. Extensive planning must precede development activities. A major step is the selection of a production facility. At the present time, the alternatives are: fixed platforms, gravity platforms, tension-legged platforms, guyed towers, and subsea systems.

Fixed platforms have evolved from the simple wooden structures used in the bayous of Lousiana in earlier days. Steel truss production structures (see Figure 2.15) are generally used for offshore petroleum production. The platform is permanently attached to the seafloor by steel pilings and supports one or more decks on which drilling and production equipment is mounted. For both drilling and production,

Figure 2.15. Offshore steel production platform. Photo courtesy U.S. Geological Survey.

a large amount of equipment has to fit on a very compact platform, often causing complications.

Fixed platforms have been constructed for increasingly deeper waters. The largest thus far is the Cognac platform installed in 1,025 feet of water in the Gulf of Mexico. Many production wells are drilled directionally into various portions of a field from the platforms (see Figure 2.16). The Beta field in San Pedro Bay off California has the densest spacing of any offshore field, with one well per fifteen acres. The Gilda platform in the Santa Barbara Channel has slots for 96 production wells, 16 more than the previous record holder, Platform Ellen in San Pedro Bay. The deepest producing offshore well is in Block 25 Grand Isle off Louisiana, where production is from 20,500 feet below

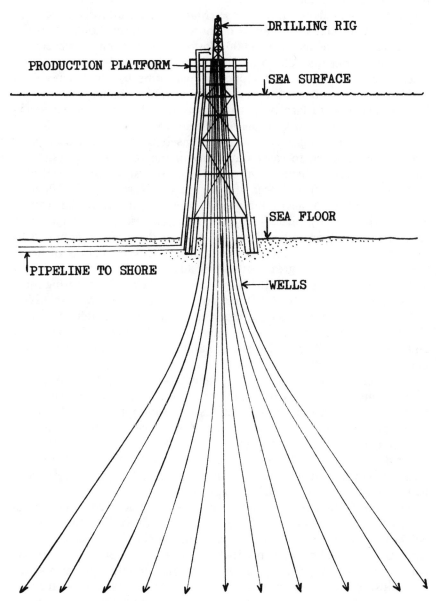

Figure 2.16. Diagram of directional drilling from an offshore production platform.

the seafloor. There is a limiting water depth below which conventional fixed platforms cannot be used, the two major technical problems being foundations and dynamic response. Many fixed platforms have been constructed in relatively shallow offshore waters throughout the world. In the Gulf of Mexico over 3,000 platforms have been emplaced.[21]

The tension-legged platform is designed for use in deeper waters where conventional bottom-supported platforms become increasingly expensive. It is similar to a taut-moored buoy, being a buoyant structure held beneath the surface by tension members. The buoyant structure, in turn, supports a working platform held above the water surface by means of vertical columns that are themselves buoyant. This type of platform is designed to carry out many of the activities necessary for offshore operations. Installation of the world's first tension-legged platform is underway in 485 feet of water in the North Sea. It will have slots for 32 wells. A steel template has been positioned on the seafloor and will be held by piles. The hull and deck resemble a large semi-submersible drill rig but will be attached to the seafloor template by four tensional mooring lines at each corner. In water depths of under 500 feet there is no great economic advantage to a tension-legged platform, but by extending the length of the hollow steel mooring lines, fields in water depths to 2,000 feet could be developed with relatively little increase in cost. Because mooring lines remain under tension at all times, the platform does not move up and down with the waves and the distance between the platform and the seafloor remains virtually unchanged.

The primary advantage of the tension-legged platform over the conventional platform is that, in deep water, it requires less steel and thus costs less. Other important features are its mobility and its flexibility with regard to depth and location. It can be relocated with comparative ease, and it need not be designed and built for a specific location or water depth. With modification it can be made suitable for almost any location within a general region, allowing construction to begin prior to identification of exact field location.

The world's first commercial guyed tower is under construction for use in the Gulf of Mexico. When completed, the 1,300-foot structure will be located in over 1,000 feet of water and will be able to accommodate simultaneous drilling and production operations. Guyed towers are considerably less expensive than conventional structures in deep water, with most of the savings in reduced steel requirements and construction expenses. The guyed tower is a trussed structure that rests on the seafloor without pilings and is held in place by guy lines. The tower is designed to sway between one and two degrees during the passage of large waves, so the well conductors must flex at the tower base. The

47,000-ton Gulf of Mexico structure will be held in place by twenty guy lines arranged symmetrically around the tower and attached to clump weights on the ocean floor. The weights are designed to be lifted off the bottom only by large storm waves, thus softening the mooring system and allowing the tower to displace with the wave. Beyond the clump weights, the guy lines are attached to anchor piles. If a line should fail, the structure will not be in danger of collapse, as the guying system is designed to be highly redundant.[22]

The use of gravity or pileless platforms is one means of installing huge offshore structures in deep water under difficult weather and sea-bottom conditions. There are a variety of gravity platform designs, but most are made primarily of concrete. Concrete is preferred at present for the North Sea, where most of the gravity platforms are located. Steel gravity structures have been designed for areas where a hard sea bottom makes pile driving very difficult (such as offshore from the Congo).

As their name implies, gravity platforms rest on the seafloor, stabilized by their own weight without deep pilings. The Ninian platform, in the United Kingdom section of the North Sea, weighs 600,000 tons. The principal technical requirement for the stability of such platforms is the prevention of foundation failure, which can be caused by: sliding between the base of the gravity structure and the soil, bearing capacity failure, softening along the rim of the base, and liquefaction of the soil.[23]

Concrete is a suitable material for a number of reasons, including ease of construction and resistance to corrosion and fire. In the North Sea, unlike the Gulf of Mexico where deep deposits of soft clay predominate, the seafloor consists of stiff clays and dense sands able to support the heavy loads introduced by the concrete platforms. A difficulty associated with gravity platforms is the scarcity of coastal sites in which they can be built. Unlike conventional platforms, gravity platforms are constructed in an upright position and completed largely onshore before being towed vertically to their destination at sea. The platform fabrication site must have very deep water and a clear path out to sea with continuous depths of as much as 600 feet. Few coastal sites meet these requirements. The concrete platforms in the North Sea stand in water of about 325 to 500 feet deep, but gravity structures for water depths of 650 to 1,000 feet are under consideration. These structures would be constructed and installed in a manner similar to those already in existence, but they will require very deep inshore waters for fabrication and tow-out. It is possible that the deep and sheltered Norwegian fjords could be used to build such large gravity platforms. Because the most important characteristic of concrete is its ability to

absorb tremendous compressive stress with relatively inexpensive construction, the cost increase in deeper water is not as great as the cost
increase for the larger steel platforms.[24]

Conditions must be carefully analyzed before the installation of a
gravity platform. The sea bottom where the platform is to be set is
critical, as any last-minute change in location can affect the entire design
of the structure. It is also difficult to position a large structure exactly
on the spot where the soil samples were taken. It appears possible,
however, that a combination of the advantages of both conventional
steel structures and gravity-type platforms might be achieved in a single
structure. Short piles to provide added stability to a gravity platform
might make the most of both approaches.[25]

Subsea completions involve placing wellheads on the ocean floor
rather than on production platforms. The produced oil or gas is
transferred from the subsea wellhead either to a nearby platform or to
a shore facility for processing. There are many subsea completions in
the offshore areas of the world, the deepest being in 600 feet of water
in the Enchova field off Brazil. Eight subsea completions are in progress
in the Tazerka field off Tunisia in 820 feet of water, which will be a
new water depth record. Several subsea completion systems are available
for use in fields that do not lend themselves to conventional platform
development because of either limited petroleum reserves or deep water.
Complete subsea systems are expensive, as all development wells have
to be drilled from mobile rigs and seafloor gathering equipment is
complex and costly, but in water depths in excess of 1,000 feet such
complete systems may become economic when compared to giant
platforms.

No two subsea completion systems are alike; each field has unique
problems that must be solved with unique approaches. There are a
variety of configurations including wet systems, in which the wellhead
is exposed to the water, and dry systems, which contain essentially
conventional wellhead equipment within watertight chambers at atmospheric pressure. The chambers can contain air, allowing men to
work on the wellhead equipment, or (when not occupied) a nitrogen
atmosphere, to prevent fires and explosions. The wellhead control
equipment located on the seafloor is freed from vulnerability to damage
from ships and storms and is designed to shut down in response to
production problems within the well. Sometimes platforms are spaced
at distances that do not permit full recovery of petroleum from an
offshore field. In this case subsea systems are used to produce from
field areas not reached by the platforms.

For offshore production from areas in iceberg lanes, fixed platforms
can be built to withstand collisions with one-million-ton icebergs but

would not survive collisions with larger ones. Thus, floating production systems, which may be disconnected in the event of the approach of an iceberg or the accumulation of pack ice, are under consideration. Since such systems have been successfully installed in the North Sea, the technology has been proved. Its adaptation to an area of iceberg occurrence has been suggested for the Hibernia field, offshore from eastern Canada. The wellheads would be recessed to a depth of 25 feet below the seabed to prevent damage due to iceberg scour. The crude oil would be routed to the floating production unit on the surface, which would be designed to accommodate 20 producing wells.[26]

Oil is transported from production wells on the outer continental shelves to shore either by pipelines or by bulk carriers, such as tankers or barges. Currently almost all offshore gas is transported to shore by pipelines, but shuttle barges have been designed to transport gas from offshore fields that are too small to support a pipeline and too large to waste. In proposed barge transport of offshore gas, natural gas would be taken from a wellhead site and loaded through a barge mooring system to a shuttle barge under pressures of about 2,500 pounds per square inch. To off-load the barge at an existing platform with a pipeline connection or at the shore, this pressure would be bled down to line pressure and the remaining gas scavenged by compression. Once a reservoir is depleted or proves large enough to justify a pipeline, the shuttle barge system could be moved to another site and reused, unlike a pipeline, which cannot be recovered. There are a large number of shut-in wells in the Gulf of Mexico, many of which are marginal fields. By minimizing the production facility, the economics of transporting compressed gas by barge may make production of marginal or early-developing reservoirs possible.[27]

Barges and tankers are used as a temporary means of transportation of oil during field development and also to transport oil from low-production fields. Usually, however, plans for new outer–continental-shelf petroleum development include pipelining in one form or another. Offshore pipelines serve two major purposes: the gathering of the gas or oil and the transmitting of them to shore. Gathering lines move production fluids to a central point for measuring, storage, or treatment and then terminate at a final metering point. Pipelines that move oil and gas beyond this point are known as transmission lines.

There are three primary methods to lay pipe offshore. The most common is the lay barge or "stovepipe" technique in which sections of pipe, usually coated with concrete, are welded together on a lay barge (see Figure 2.17) and released into the water as the barge moves forward. Lay barges can be used for pipe from 4 to 52 inches in diameter. A second technique is the reel barge method, in which long

Figure 2.17. Lay barge for offshore pipelines. Photo courtesy U.S. Geological Survey.

sections of pipe are welded on land, wound onto a large reel on the barge, and later laid directly from the reel into the sea. For smaller-diameter pipe, reel barges are often more economical than lay barges. A third method is to pull pipe from fabrication facilities onshore into the water. Although pipe was once laid directly on the seafloor, it is now more common to bury offshore pipelines to avoid damage from currents, storm waves, and anchors. A burial barge is used to sink the pipe beneath the seabed surface, usually by displacing soil with a high-pressure jet.

Future use of offshore pipelines depends upon the two basic problems of water depth: diver and structural limitations. While working in deeper water, lay barges use an articulated structure with adjustable buoyancy, known as a stringer, to support the pipe between the barge and the seafloor. Another approach for increased water depths is the use of an inclined or vertical assembly area for the pipe, which tends to reduce the overbend when the pipe is no longer supported by the stringer. Methods of laying pipe in deep water require increased capabilities in navigation and positioning. Thicker-walled pipe is necessary to withstand the combination loads of bending and high external pressures of the deep-water environment.

A 20-inch pipeline has been laid from Algeria across the Mediterranean Sea to Sicily. At some parts of the route the sea is over 2,000 feet deep. Pipelines have also been designed to be installed under the ice in Arctic areas. They are usually laid in a trench cut by an underwater trenching plow near the shore to protect the pipe from ice scour in the shallower waters. The pipeline is pulled from onshore by a pull winch located on the ice surface. Such techniques will be utilized to produce the numerous subsea wells off the Arctic Islands of Canada.

Summary

The exploration for petroleum has become a complex and highly sophisticated endeavor, but actual drilling is still required to verify the existence of a commercial deposit. The earliest known indication of a petroleum deposit was an oil seep and such surface manifestations led to many discoveries. Next, the field mapping of geologic structures was utilized, with the recognition of the characteristic oval-shaped outcrop pattern of an anticline being a guide to favorable drilling locations. Later, subsurface methods of exploration came to be used where the bed rock was not adequately exposed. The most elaborate and expensive, but most effective, of these methods is the seismic survey during which shock (or seismic) waves are reflected from a number of subsurface formations and the length of time required for a round trip measured to create a vertical picture of the arrangement of the subsurface strata. The reflection seismograph has become almost a precision instrument allowing the identification of relatively minor subsurface stratigraphic features. However, it still does not provide a direct measurement of the presence of a commercial accumulation of petroleum. A direct verification requires a hole in the ground and rotary drilling is the most common method of providing such a hole. It is a hydraulic process in which a hollow column of steel drill pipe, attached to a hard-toothed bit, is rotated from a surface rig. During drilling, a stream of drilling mud is in constant circulation down the pipe and out through the bit to remove the rock cuttings and to counterbalance the formation pressures, thus helping to prevent blowouts.

Drilling methods offshore are similar to those employed onshore, except that marine risers have been developed to conduct the drill string from the surface rigs to the hole being drilled and also to conduct the drilling mud back up to the rig. They are designed to permit some lateral and vertical movements during drilling operations without break-ing off the drill string.

There are four basic types of offshore exploratory drilling platforms: barges, drill ships, jack-ups, and semisubmersibles. Barges and drill ships are simply barges and ships designed to accommodate drilling rigs. Barges are not self-propelled and are limited by weather conditions, having poor motion characteristics. Drill ships have the same general lines as merchant ships and are self-propelled. Drill-ship positioning is often accomplished by a dynamic positioning system, a series of propellers or thrusters coupled to sensors that detect and compensate for movement. A drill ship has established the world water-depth record for conventional petroleum exploration by drilling a 20,023-foot hole in 4,876 feet of water off Newfoundland.

Jack-up rigs are drilling platforms with legs that can be moved up and down. When the legs are extended the platform can elevate itself above the ocean surface and temporarily become a bottom-standing platform. By retracting its legs, it becomes a floater and can be towed to a new location. Jack-ups can drill in water up to about 350 feet deep and are the most popular of all offshore drilling units because of their relatively low cost. The semisubmersible is the newest type of offshore rig and is engineered for severe weather conditions. These rigs have a platform deck supported by columns connected to large underwater displacement hulls that are flooded on site to reduce wave-induced motion by submerging the major buoyancy members below the level of wave action. Several large semisubmersibles have been designed to operate in 3,000 to 4,000 feet of water. Their primary disadvantage is high cost.

If commercial accumulations of petroleum are discovered offshore, the development phase begins. A major step is the selection of a production facility. The three main types of offshore production systems currently in use are the fixed platform, the gravity platform, and subsea completions. Other possible choices include tension-legged platforms and guyed towers. Fixed steel platforms have been constructed for increasingly deeper waters. The largest tower to date was installed in 1,025 feet of water in the Gulf of Mexico and is one of over 3,000 offshore platforms that have been emplaced in that region. The use of gravity or pileless platforms made of concrete is a means of installing huge offshore structures in deep water under difficult weather conditions. The Ninian platform in the North Sea weighs 600,000 tons. Concrete is a suitable material for a number of reasons, including ease of construction and resistance to corrosion and fire. Subsea completions involve placing wellheads on the seafloor rather than on production platforms. There are many subsea completions in the offshore regions of the world, the deepest being in 600 feet of water off Brazil. They are used for smaller fields or for fields in deep waters where platform construction is not economical.

Arctic regions, which contain some of the most promising prospective petroleum areas remaining in the world, have specific problems relating to the intense cold. Icebergs can endanger offshore drilling units. Either they must be towed out of the drilling area or the riser must be disconnected and the drilling unit moved. Floating production systems, which may be disconnected in the event of the approach of an iceberg or the accumulation of pack ice, are under consideration. Offshore ice platforms have been constructed for Arctic drilling.

Oil is transported from offshore production wells either by pipelines or by bulk carriers, such as tankers or barges. Currently, almost all offshore natural gas is transported to shore by pipelines. The most common method of laying pipe offshore is with a lay barge on which sections of pipe, usually coated with concrete, are welded together and released into the water as the barge moves forward. A burial barge can be used to sink the pipe beneath the seabed surface, usually by displacing the soil with a high-pressure jet. Future use of offshore pipelines depend upon the problems of water depth, including diver and structural limitations. Even so, pipelines have been successfully laid in water more than 2,000 feet deep.

Primary recovery averages around 20 percent of the in-place oil in a field. However, in the United States average primary recovery has been somewhat higher, around 28 percent, due to the exceptional reservoirs in East Texas and southern Louisiana. The amount of in-place nonassociated natural gas recoverable during the economic life of a reservoir is quite high, about 75 to 80 percent. Gas wells often flow, but oil wells usually require an artificial lifting system. Currently most of the oil lifting is done by some variety of the "horse head" walking-beam pumping system.

Secondary recovery methods (usually waterflooding) can improve recovery to between 30 to 50 percent of the in-place oil. Modern waterflooding attempts to maintain or restore reservoir pressure by the injection of water around the periphery of a field so that it displaces the oil upward in a manner similar to a natural water drive. With continued production, the percentage of water produced (water cut) gradually increases until the field is no longer economical to produce. The domestic oil industry recovers 6 to 7 barrels of water for each barrel of oil. There are techniques available to recover additional in-place oil from a reservoir after the termination of a waterflood project. These are known as enhanced oil recovery methods.

Notes

1. Carl H. Savit, "Geophysics Will Find the Elusive Strat. Trap," *Oil and Gas Journal* (November 17, 1980), pp. 59–62.

2. Allen L. Hammond, "Bright Spot: Better Seismological Indicators of Gas and Oil," *Science* (August 9, 1974), p. 516.

3. J. W. Jenner, "Drilling for Oil," in *Modern Petroleum Technology,* G. D. Hobson, ed. (New York and Toronto: John Wiley & Sons, 1973), pp. 108–144.

4. *Ibid.*

5. Charles E. Chadwick, "Current Deep-drilling Technology and Limitations," *Oil and Gas Journal* (October 19, 1981), pp. 251–256.

6. "Giant Land Rigs Start Deep Drilling," *Oil and Gas Journal* (January 5, 1981), p. 80.

7. Chadwick, "Current Deep-drilling Technology."

8. Jenner, "Drilling for Oil."

9. Don E. Kash, et al., *Energy Under the Oceans* (Norman: University of Oklahoma Press, 1973), p. 37.

10. G. W. Leonhardt, "Drilling in Record Water Depth Was an Operational Success," *World Oil* (February 1980), pp. 57–60.

11. W. D. Moore, III, "Offshore Drilling Rig Building Record Set," *Oil and Gas Journal* (May 4, 1981), pp. 123–126.

12. Olivier Pierrat, "More Technology is Needed to Drill in 10,000 Feet of Water," *World Oil* (April 1980), pp. 54–56.

13. "Sea Exacts Its Price in Hunt for Oil," *U.S. News & World Report* (March 1, 1982), p. 6.

14. G. L. Hood, H. J. Strain, and D. J. Baudais, "Offshore Drilling From Ice Platforms," *Journal of Petroleum Technology* (April 1976), pp. 379–384.

15. Joseph P. Riva, Jr., *Secondary and Tertiary Recovery of Oil* (Report prepared for the Subcommittee on Energy, House Committee on Science and Astronautics [Committee Print 40-382], U.S. Congress, Washington, D.C., October 1974).

16. T. M. Doscher and F. A. Wise, "Enhanced Crude Oil Recovery Potential—An Estimate," *Journal of Petroleum Technology* (May 1976), pp. 577–578.

17. R. F. Krueger, "Advances in Well Completion and Stimulation During JPT's First Quarter Century," *Journal of Petroleum Technology* (December 1973), p. 1459.

18. Bill D. Berger and Kenneth E. Anderson, *Modern Petroleum* (Tulsa, Oklahoma: Petroleum Publishing Company, 1978), pp. 142–160.

19. Todd M. Doscher, "Enhanced Recovery of Crude Oil," *American Scientist* (March-April 1981), pp. 193–199.

20. *Ibid.*

21. "Gulf of Mexico Report: Records and Achievements in the U.S. Sector," *Ocean Industry* (November 1981), pp. 57–62.

22. Dan McNabb, "Guyed-Tower Platform Design Nearing Offshore Test in Gulf of Mexico," *Oil and Gas Journal* (July 14, 1975), p. 86.

23. Ivar Foss, "Concrete Gravity Structures for the North Sea," *Ocean Industry* (August 1974), p. 58.

24. Finn Rosendahl, "Concrete Structures Offer New Options," *Offshore* (September 1979), p. 61.

25. John L. Kennedy, "New Types of Gravity Structures Near Completion," *Oil and Gas Journal* (May 5, 1975), pp. 210–220.

26. Leonard LeBlanc, "Producing in Iceberg Lanes," *Offshore* (October 1980), pp. 56–62.

27. Donald W. Fowler, "Transport Process Pegged For Small Gas Fields," *Offshore* (July 1981), pp. 108–110.

3
Enhanced Oil Recovery

If enhanced recovery processes are not applied to old oil fields before they are abandoned, it is likely that their remaining oil will be forever lost. After secondary recovery methods have been used to their maximum potential, the extraction of additional oil will require the application of more sophisticated and expensive techniques. These techniques usually attempt to reduce oil viscosity and capillarity by the introduction of heat (steam) or of other injected substances such as carbon dioxide, polymers, solvents, surfactants, and micellar fluids in various combinations depending upon reservoir conditions and crude oil properties. Enhanced oil recovery processes (sometimes termed tertiary techniques) can further increase recovery to 40 to 80 percent of the total in-place reservoir oil, depending upon the process employed and upon the physical properties of the reservoir and of the in-place oil.

Thermal Methods of Recovery

Thermal recovery methods are used for the production of heavy crude oils, the recovery of which is impeded by viscous resistance to flow at reservoir temperatures. The heating of heavy crude oils markedly improves their mobility and the efficiency of their recovery. Heat may be introduced into the heavy oil reservoir by injecting a hot fluid, such as steam or hot water, by burning some of the heavy oil in place in the reservoir, or, rarely, by electric heating.

Crude oils are generally classified, on the basis of relative mobility, into tar sands, heavy crude oils, and medium and light crude oils. Tar sands contain immobile oil, which will not flow into a well bore even if stimulated by enhanced oil recovery processes. Heavy crudes are considered to be those with enough mobility that, given time, they will be producible through a well bore in response to the use of enhanced recovery methods. Crude oils below 20 degrees API gravity are usually considered heavy, while crudes with API gravities between 20 and 25 degrees are considered medium. The API gravity scale was developed by the American Petroleum Institute to express the density of liquid petroleum. According to this scale, water has a gravity of 10 degrees API and liquids lighter than water, such as petroleum, have API gravities numerically greater than 10. Although the upper limit placed on heavy oils is somewhat arbitrary, there does appear to be a significant breaking point at about 20 degrees API. In general, conventional waterflooding is most effective with light crude oils of 25 degrees gravity and above and becomes progressively less effective with oils below 25 degrees. With 20-degree and below oil, waterfloods are essentially ineffective and thermal recovery becomes necessary. Most thermal recovery projects, such as steam injection, take place in reservoirs that contain heavy oil between about 10 and 20 degrees API gravity. Very few steam projects are successful in recovering oil that is below 10 degrees API gravity.

Heavy oil is formed when lighter oils move upward toward the earth's surface and encounter descending meteoric waters containing oxygen and bacteria. At temperatures below 200 degrees Fahrenheit degradation begins. A tar-like material is formed at the oil-water contact and eventually invades the whole oil accumulation. Such water washing and bacterial degradation of mobile, medium-gravity crude oils results in the formation of heavy oils. Water washing removes the more water-soluble, light hydrocarbons, especially the aromatics, and biodegradation preferentially removes the normal paraffins, resulting in an increase in density and sulfur content.[1] The nature of individual heavy oil deposits varies widely, and deposits seldom remain chemically homogeneous for horizontal distances of more than 50 feet.

It appears that the amount of heavy crude oil in the world may be about equal to the amount of lighter crude oil, although the former occurs mostly in the Western Hemisphere and the latter mostly in the Eastern Hemisphere. The heavier crudes are a much less desirable energy resource than the lighter crudes because they are much more costly to extract and to process. They have not been exploited to the degree of the lighter crudes, however, so they remain available for development. Heavy oil development becomes more necessary as the supply of the lighter crudes declines, causing oil prices to rise and shortages to develop.

Steam Soaking and Steam Flooding

In the United States, the use of steam to recover heavy oil began in California about twenty years ago. The most common method of steam treatment is the steam soak (also termed cyclic steam injection or huff-and-puff). This is essentially a well-bore stimulation technique in which steam is injected into a producing well for a number of weeks, after which the well is shut in for several days before being put back into production, often with a significant increase in output. It is sometimes economic to steam soak the same well several times, even though oil recovery generally declines with each succeeding stimulation.[2] Some wells have been reported to have had as many as sixteen cycles of steam soaking, followed by the production of heavy oil.

A steam soak is economically efficient only in permeable reservoirs containing more than 1,000 barrels of oil per acre-foot and measuring over 50 feet thick so that vertical (gravity) drainage can occur. There are only a few large domestic heavy oil reservoirs that meet these criteria and they are located in California.[3]

As cyclic steam injection is primarily a well-bore stimulation process, continuous steam injection, which contacts a greater portion of the reservoir, appears necessary to achieve the most efficient heavy oil recoveries. Since 1968, there has been a large increase in continuous steam injection in California heavy oil fields. Continuous steam injection (steam flooding) is a displacement process, similar to waterflooding. The steam is pumped into injection wells (sometimes after the wells have been fractured to increase permeability) and the oil is displaced to producing wells as in conventional flooding operations. Because of the relatively high cost of steam, water is sometimes injected at an optimum time to push the steam toward the producing wells. Recoveries of up to 30 to 40 percent (with a maximum of 70 percent under optimal conditions) have been reported for successful steam floods.

During steam flooding, the steam flows over the oil, transferring heat by conduction. The oil, with its viscosity reduced by the heat, is conveyed by the steam to the producing well. Since the steam serves two functions, the heating and the transporting of the oil, some steam must always be circulated through the formation without condensing. Even in some of the most favorable reservoirs, it is necessary to consume an amount of energy equivalent to burning 25 to 35 percent of the crude oil produced to generate the required amount of steam. This operation is limited to reservoirs that contain over 500 barrels of heavy oil per acre-foot, as it requires about that much fuel to generate the steam required to effectively heat an acre-foot of reservoir rock to the necessary recovery temperature.[4]

Some operators are experimenting with the injection of a chemical foam into the reservoir to plug channels and thus prevent the waste of steam. There is danger, however, that the foam could plug the formation. In another steam process currently being tested, the heat is provided by injecting steam from beneath the reservoir by using horizontal holes directionally drilled in a radial pattern from a large-diameter central well. The heat from the steam injected into this system will be distributed into the overlying reservoir to mobilize the heavy crude oil. This technique takes advantage of the tendency of steam to rise. Gravity carries the flowing oil (and formation water) downward into the horizontal holes for collection. Laboratory tests have indicated that recoveries of up to 65 percent of the in-place oil could result from this process.[5]

The heat loss that occurs when steam is injected thousands of feet down an injection well is a serious problem for heavy oil producers. Bore-hole steam generators are under development to counter down-hole heat loss by replacing the usual 20-by-40-foot surface boiler with a more efficient combustion chamber small enough to enter the 7-inch bore hole of an injection well. The elimination of the surface boiler also aids in the problem of conforming to air pollution standards. These generators have been effective in tests but are not ready for commercial use. Some models incorporate the injection of combustion gases as well as steam into the reservoir. The addition of combustion gases to the reservoir improves heavy oil recovery because oil viscosity is a function of temperature as well as the amount of carbon dioxide in solution in the oil. Research is also under way on processes in which combustion gases are superheated above ground and pumped down injection wells in a closed system.

In-Situ Combustion or Fire Flooding

The mechanics of crude oil displacement in an in-situ combustion operation is similar to that in a steam flood, except that the steam is generated by the vaporization of reservoir formation water or of injected water. In fire flooding a heavy oil reservoir, the in-place oil is ignited and the burning front moved along by continuous air injection. There are three variations of the process: forward combustion, in which air is injected into a well, advancing the burning front and heating and displacing oil and water to surrounding production wells; reverse combustion, in which a short-term forward burn is initiated by air injection into a well that will eventually produce oil, after which the air injection is switched to adjacent wells (for oil so viscous that it could not move through a cold zone in a forward combustion process); and wet or quenched combustion (COFCAW—combination of forward combustion and waterflooding), a modified form of forward combustion incorporating the injection of cold water along with the air to recover some of the heat that remains behind the combustion front.[6] The air requirement is lower with water injection and, at a constant air injection rate, in-place oil may be produced faster because of the more rapid movement of the combustion zone, the increased utilization of the available energy, and the increased volume of injected fluids. The air-water combination minimizes the amount of air injected and the amount of in-place oil burned (to between 5 and 10 percent), and also improves sweep efficiency.

The costs associated with the development of heat within a heavy oil reservoir and the success of the recovery operation are influenced by the depth of the reservoir. In general, shallower reservoirs are considered candidates for steam soaks and floods, and deeper reservoirs for in-situ combustion. However, severe operating problems have been encountered in many combustion operations, hampering widespread application of this process. The problems include production well corrosion, due to the presence of hot gases, and the production of very tight emulsions of water in oil, which are difficult to treat. Thus the cost of developing heat within even relatively deep reservoirs by in-situ combustion is not always less than the cost of heating by steam injection.

Electric Heating

Nonselective electric heating techniques have occasionally been used to recover heavy oil. The process can be used along with a waterflood when the injected water is salty. Electrodes can be installed in injection wells so that an electrical current flows through the in-place oil, which acts as a resistance heating element converting the electrical energy to thermal energy. Because the current density is highest near the injection wells, however, most of the heating usually occurs within a few feet of the electrodes. It is generally more expensive to heat a reservoir with electricity than with steam, so electric heating methods have been suggested for areas where other thermal recovery techniques cannot be employed because of air pollution problems.

An improved electrical technique called selective reservoir heating has been proposed. In this method, the current is applied selectively to those parts of the reservoir that have been bypassed by a waterflood or other recovery process. Concentration of heat in the desired portions of the reservoir is achieved by establishing a current flow in the reservoir along paths that differ from the paths of oil flow. Electrodes may be installed in adjacent injection wells with alternate electrical polarity. This will cause the current to flow from well to well through the unswept portions of the reservoir. The injection of fresh water before electric heating will create high resistivity in the more permeable zones, causing the current to flow through the less resistive (less permeable) parts of the reservoir that are in need of heating to increase in-place oil movement. [7] To avoid excessive heating near the electrodes during this process, high-salinity water can be injected while the electricity is applied.

Mathematical simulation appears to be the best method available to predict the response of a reservoir to selective heating, but thus far these newer electrical techniques have not been field tested.

Chemical Methods of Recovery

Some reservoirs, because of differences in the physical characteristics of either the rock or the oil, do not respond well to waterflooding. In certain of these reservoirs oil recovery in response to water injection can be improved by the addition of chemicals to the water.

Polymer Waterflooding

Oil recovery by water injection can sometimes be improved by increasing the viscosity of the injected water. Polymers, either polysaccharides or polyacrylamides, which are large organic molecules capable of increasing the viscosity of water, have application in reservoirs containing high viscosity oil, which ordinary water would mostly bypass. Polymers, however, can also plug the formation and retard subsequent water injection. To overcome this tendency, water-soluble cationic and anionic polymers have been developed to be injected as mixtures designed to move in a bank through the zones of highest permeability in the reservoir. Because the polymer mixture reduces permeability in high–sweep-efficiency areas, subsequent injection water is diverted to less permeable areas. As the injection water encounters the polymer bank, which is miscible in water, the water viscosity increases, also improving overall sweep efficiency.[8]

Surfactant Systems

Another method of increasing oil recovery above conventional water-flooding levels utilizes surfactants (detergents) in various combinations. The microscopic oil displacement efficiency of a conventional waterflood is limited by the magnitude of the capillary forces existing in the pore spaces of the reservoir. The injection of the surface-active surfactants in aqueous solution affects interfacial tension and reduces capillary pressure, permitting the release of immobilized oil. To achieve such a release, the interfacial tension between oil and water must be reduced several orders of magnitude below normal values. The recovery of a substantial amount of the oil remaining in regions where the capillary forces are greater than the viscous forces can be achieved only by a reduction in capillary forces relative to the forces exerted by the flowing fluids. In crude oil reservoirs, however, the effectiveness of simple surfactants is reduced by the adsorption of the surfactants into the reservoir minerals, the salting-out by reservoir brines, and bacterial degradation. Also, and even more importantly, a reduced interfacial tension prevents the fine reservoir pores from taking up water and permits the surfactant solution to finger quite rapidly to the production wells, bypassing too much of the in-place oil.[9]

To overcome such problems, surfactant systems are optimized to involve typically at least five components: water, oil, surfactant, co-surfactant, and salt. The solutions are also frequently loaded with water-soluble polymers, hydrolyzed acrylate polymers, and polysaccharides to prevent fingering and also to improve reservoir sweep efficiency. Such systems are called micellar-polymer fluids because the surfactant normally is present in amounts above the critical concentration for micelle (microemulsion) formation.

Surfactants, mixed at critical concentrations with water, cling together to form micelles or microemulsions. If such micellar systems are injected into an oil reservoir, the micelles can solubilize or dissolve the oil by the entrapment of tiny oil droplets into their centers, thus causing miscibility. The mixture is a microemulsion that causes the displacement of the contacted in-place oil.

The high cost and unique characteristics of micellar-polymer fluids make it necessary to control carefully their quality and distribution in the reservoir. The process appears to require more sophisticated control than is usually achieved in an oil field. Although very encouraging results have been obtained from laboratory studies, field results thus far have been generally discouraging.[10] Because each reservoir has unique properties, specific surfactant systems must be designed for each application. Surfactant flooding continues to receive widespread attention both in laboratory and in field research, however, as the promise of the method is so great. Unfortunately, the increasingly high costs of surfactants and polymers, plus the discouraging field results, have caused those companies with the most experience in surfactant flooding to reduce future plans for employing the method.[11]

Caustic Flooding

Caustic or alkaline flooding involves the use of strongly alkaline water solutions of such chemicals as sodium hydroxide, sodium silicate, and sodium carbonate to react with the constituents present in some crude oils or present at the reservoir rock–crude oil interface to form detergent-like materials (surfactants) that reduce the ability of the formation to retain oil.[12] The surfactants are formed from the neu-tralization of petroleum acids by the alkaline chemicals in the displacing fluids. Since the content of such natural petroleum acids is usually higher in the heavier crudes, this process appears to be applicable to the recovery of only moderately viscous, rather heavy, napthenic type oils.[13] The cost of caustic flooding is considerably less than the cost of micellar-polymer flooding. Some estimates indicate that caustic recovery will cost only about one-third as much as micellar-polymer recovery.[14]

Gas Methods of Recovery

Carbon Dioxide

Carbon dioxide is capable of miscibly displacing some oils and therefore permits the recovery of most of the in-place oil that it contacts. The miscible displacement overcomes the capillary forces that act to retain in-place oil in the pores of the reservoir rock. However, the carbon dioxide is not initially miscible with the in-place oil. As the carbon dioxide contacts the crude oil in the reservoir, either some of the hydrocarbon constituents of the crude are vaporized and a portion of the resulting mixture becomes miscible with both the carbon dioxide and the in-place oil, or the carbon dioxide may condense in the residual oil to form a new miscible phase. Therefore, the reservoir operating pressure must be kept high enough to develop and maintain a mixture of carbon dioxide and hydrocarbons that is miscible with the in-place oil at reservoir temperature. During the later stages of injection, the carbon dioxide may be driven through the reservoir by lower-cost inert gas or water. To achieve a higher sweep efficiency, carbon dioxide and water may be injected in alternate cycles (see Figure 3.1).[15] Also, the carbon dioxide may be dissolved in water and injected as carbonated water or mixed with lesser amounts of methane.

Although field studies have demonstrated that carbon dioxide at high pressures will mobilize and displace oil that has been trapped by waterflooding, the actual mechanism by which the carbon dioxide recovers the in-place oil is not completely understood. A precise understanding of the process could lead to an optimization of carbon dioxide use, which currently consists of quite large volumes per additional barrel of crude oil recovered.[16] Studies have indicated that the low viscosity carbon dioxide tends to finger through the more viscous formation water. It is only after the carbon dioxide has displaced a significant portion of this water that it can reach the residual oil that is immobilized by the water. It dissolves in and swells the globs of oil on contact and as the swelling globs join together, the oil phase again begins to flow. The carbon dioxide has to displace the formation water before displacing the residual crude oil, making the process inefficient. To achieve the maximum displacement per volume of carbon dioxide,

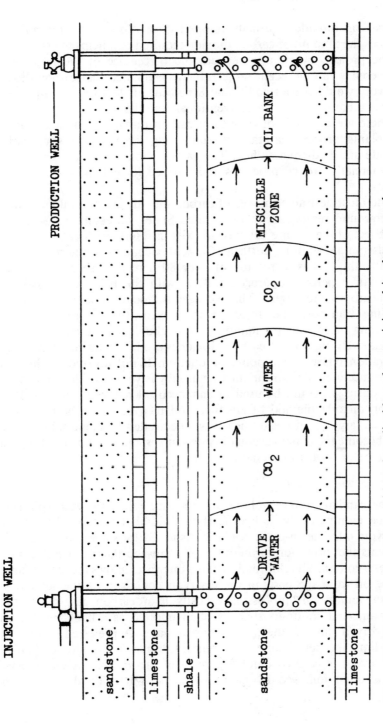

Figure 3.1. Enhanced oil recovery with carbon dioxide and water injected in alternate cycles.

the carbon dioxide is usually injected discontinuously. This not only reduces the amount of carbon dioxide required and helps the economics of the project, but it also reduces the percentage of the in-place oil recovered.[17] In one recent project, it was estimated that carbon dioxide recovery would net an additional 8 percent of the original in-place oil after the completion of a conventional waterflood.

There are a variety of sources of carbon dioxide. A number of sour gas plants process natural gas containing large amounts of carbon dioxide and hydrogen sulfide. The carbon dioxide can be removed by commercially proved processes and purified for enhanced oil recovery use.

The most desirable sources of carbon dioxide are naturally occurring underground reservoirs. In the United States, an estimated 30 trillion cubic feet of the gas are believed to exist in reservoirs located in central Mississippi, Wyoming, northeast New Mexico, and southern Colorado. Usually this gas is a mixture of over 80 percent carbon dioxide and less than 5 percent hydrogen sulfide, with the balance consisting of hydrocarbons and water.[18] The hydrogen sulfide must be removed before pipeline transport and injection to prevent the oil reservoir from becoming more sour (containing more sulfur compounds).

Relatively high-purity carbon dioxide is produced during the hydrogen generation phase of ammonia production. If not utilized in the plant to produce urea, the carbon dioxide usually is vented to the atmosphere. This waste gas could be used in enhanced oil recovery projects. If the carbon dioxide method of enhanced oil recovery is to be extensively utilized, large quantities of the gas will be needed. The natural reservoirs may become the major carbon dioxide sources, but the industrial supplies could be of local importance.

Hydrocarbon Miscible

The capillary forces holding the in-place oil in the reservoir pores can be eliminated by a hydrocarbon fluid that is miscible with the reservoir oil. The in-place crude is displaced by solvent action, which prevents the formation of the interfaces between the injected fluids and the reservoir oil. The elimination of the interfaces allows the displacement of the oil from the parts of the reservoir contacted by the solvent. As the solvent in this process is a costly hydrocarbon gas, rather than being injected continuously it is often injected as a bank or slug and then driven through the reservoir with water. A preferred solvent has been liquefied petroleum gas, primarily propane. The slug or bank of this solvent is usually about 5 percent of the volume of the reservoir and is often followed by natural gas and water. Field tests of this process have been disappointing because of the fingering of the liquefied petroleum gas through the oil bank due to the gas's much lower viscosity. Similar processes have been developed using different fluids that can mix with oil, including petroleum gases rich in ethane, butane, and pentane. But because of the high cost of such hydrocarbons and their generally short supply, they are no longer desirable injection fluids and the current trend is away from hydrocarbon miscible projects.[19]

Nitrogen and Other Inert Gases

The rising value of natural gas and petroleum gas have adversely affected the economics of enhanced oil recovery projects utilizing these now-expensive hydrocarbons. Possible less-expensive substitutes include nitrogen, boiler flue gas, and gas engine exhaust.

Nitrogen, injected under high pressure, becomes miscible with oil and can be used in enhanced oil recovery projects, especially when deep reservoirs are involved. The nitrogen is produced by cryogenic air separation. When air becomes liquid at −300 degrees Fahrenheit, the nitrogen can be separated out and is dry when leaving the cold box. It will remain inert if water is introduced, making a single distribution/injection system for both nitrogen and water feasible. Thus, nitrogen can be injected alternately with water for mobility control within the reservoir. The miscible nitrogen dissolves and removes the oil contacted from the reservoir rock.

The most significant difference between the boiler flue gas and the gas engine exhaust systems is that the flue gas process requires the use of a steam turbine and the gas engine exhaust process uses a reciprocating gas engine. The combustion of methane produces wet gases that have to be dried prior to injection to prevent the forming of a corrosive electrolyte. The exact composition of combustion gases after treatment and prior to injection varies depending upon fuel composition and process used, but the major constituents are nitrogen and carbon dioxide. Since about 12 percent of the average combustion gas is carbon dioxide, which is much more compressible than nitrogen, about 2 to 5 percent additional combustion gas must be injected into the reservoir to provide the same volume, at reservoir conditions, as pure nitrogen. The presence of carbon dioxide in combustion gas makes the mixture more soluble in both formation water and oil, requiring the injection of additional quantities of combustion gas.[20] Oil recovery is accomplished in this process by the miscible combustion gases dissolving and removing the contacted oil from the interstices of the reservoir rocks.

Microbiological Methods of Recovery

Laboratory and field studies on the recovery of oil using microbes began in the United States in the 1940s. Much of the more recent research in this area has been done in Eastern Europe and in the Soviet Union. The following mechanisms are considered instrumental in bacterial oil release in an underground reservoir: the physical dislodgment of oil by the growth of bacteria upon the reservoir rock surfaces; the microbial decomposition of oil with the formation of organic acids and gases; the dissolution of carbonate reservoir rock by these acids; a decreased specific gravity and viscosity of oil, due to the formation of the organic acids and the solution of produced carbon dioxide gas; an increase in pressure from the microbial generation of hydrogen, methane, and carbon dioxide; and the decreasing of surface tension by the biogenic surfactants formed.[21]

Microbial recovery technology uses a combination of bacteria that feed on oil or added nutrients. The bacterial action loosens viscous oil by producing surfactants (detergents), which promote flow, and carbon dioxide, which may displace some oil and pressurize the well. Microorganisms are being screened for their ability to produce desirable gases, solvents, and surfactants. Those identified as being capable of producing the greatest amount of such end products are tested for their ability to grow with each other under reservoir conditions. Another recovery possibility being investigated is that microbial cells may be used to selectively plug high-permeability regions in the oil reservoir and thus improve conventional sweep efficiency.[22] This selective plugging effect, however, introduces the related problem of the general bacterial plugging of reservoirs, resulting in the reduction of effective permeability and a subsequent decrease of oil production. Bactericides are necessary additives to many waterflood operations to prevent bacterial plugging and corrosion.

The organisms currently being studied for microbial enhanced oil recovery include, among others, strains of *Clostridium* and *Pseudomonas.* The process envisions the injection of such organisms into the reservoir along with an aqueous solution of fermentable carbohydrate and mineral nutrients. A significant problem appears to relate to the inhibition of the production of the desired metabolic end products in the presence of such salt concentrations as often exist in oil reservoirs.[23] Thus, research has indicated that only about 25 percent of oil reservoirs may be treatable by microbial recovery processes because of adverse reservoir conditions. However, available data indicate that the possibility of biosurfactant production in the reservoir is quite likely and that the active growth of microbes is possible and could conceivably be used in various recovery processes.[24] It is even possible that genetic engineering may be used to manipulate certain bacteria so that fewer, or even single, strains may produce several desirable end products tailored to the recovery requirements of certain reservoirs.

Because little is known about the future effects of injecting bacteria into oil wells, the environmental effects from microbiological enhanced oil recovery should be carefully examined before such processes become commercially available. Even if it is possible to develop strains of bacteria able to withstand the temperatures, pressures, and chemical environments found in oil reservoirs and still successfully generate useful modifying chemicals in sufficient amounts to commercially improve oil production, it is considered unlikely that such bacterial recovery will significantly influence domestic production until after 2000.[25]

Mining Methods of Recovery

Oil mining has been accomplished in various areas of the world including the United States, but there is only one currently operating domestic oil mine, located in the Powder River basin of Wyoming. In this mine a shaft has been driven directly into the reservoir sandstone from its outcrop. The main tunnel extends about 1,000 feet into the formation, and from it six holes were extended deeper into the sandstone to collect the oil. Peak production was initially about 70 barrels of oil per day, which has since declined to less than 50 barrels per day.[26]

In general, the approach envisioned to tap the oil remaining in what would otherwise be a depleted field is to drive a shaft through the reservoir and then construct tunnels radiating out from the main shaft below the oil sands. In the ceiling of these tunnels, holes would be drilled up into the reservoir. Gravity would draw the oil to the drill holes and then to the tunnels where a collection system would pump it to the surface by way of the main shaft.

To be worth mining, an oil sand would have to be more than 10 feet thick; otherwise there will be no gravity drive. The flow into the tunnels is determined by the permeability of the reservoir and the viscosity and specific gravity of the oil. The heavier oils are more likely to flow down into a mine rather than up into a drill hole. Depth would, of course, be a factor. Oil mines are feasible only to depths of about 4,000 feet; below this depth the temperatures become too high for miners to work.[27] There have been suggestions that oil mining may be economical and that as many as half of the depleted fields in the United States may be minable. However, domestic oil mining has not yet proved to be commercially viable, and it will be several years before even small-scale operation is tested.

The largest oil mining project currently under way is in the Yarega field of the Soviet Union. Mine shafts were dug about 500 feet deep and from these main shafts inclined shafts were sunk to the bottom of the oil sands. At the end of each inclined shaft, galleries 24 feet in diameter were constructed. Then 200 to 250 holes were drilled, radiated out from each gallery through the oil-bearing rock. Each gallery drains between four and five acres of the reservoir. The Soviets report production in excess of 1 million barrels annually from this mine and are attempting to increase production.[28]

Constraints on the Utilization of Enhanced Oil Recovery Methods

Technical Constraints

The physical and chemical changes that occur in the reservoir during enhanced recovery operations are extremely complex and not well understood. The simulation models used to predict reservoir performance and establish relative economics are usually not considered reliable.[29]

The use of surfactants to displace crude oil in the reservoir is the least-proved enhanced recovery process. Oil displacement with surfactants and micellar solutions is a miscible displacement and should, with adequate mobility control, displace all contacted oil. Although the theoretical possibilities and lab results have been very promising, field tests have been disappointing due to inadequate understanding of key reservoir parameters that control ultimate performance. The mineralogy of the reservoir has an important relationship to the chemical alteration of the injected fluids. The surfactant chemicals are highly susceptible to degradation. Improved formulations that would increase the tolerance of these chemicals to oil-field handling and adverse reservoir environments (such as high salinity) without impairing their recovery effectiveness would be a significant advance in chemical recovery technology.[30]

Field results have been much poorer than had been anticipated from laboratory studies, so it is possible that laboratory models do not well represent actual reservoir conditions. Because each oil reservoir is different, such modeling must become more site specific and therefore will become more complex as enhanced oil recovery processes are tailored to each individual field application. In chemical flood tests the reservoir seems to take over the chemical processes and the desired sweeps are not achieved.

Carbon dioxide injection is also a miscible displacement process. Injected carbon dioxide becomes miscible with in-place oil at reservoir conditions and, therefore, theoretically should recover all the oil in the parts of the reservoir contacted. However, a number of problems need to be resolved. Commercial performance cannot be determined in advance because the dynamic phase behavior of the carbon dioxide-oil–water system is not sufficiently understood. In addition, the type and nature of the displacement mechanisms existing under various reservoir conditions are not fully understood. Cost/benefit relationships of different injection methods cannot be assessed because of a lack of development of methods of mobility control. [31]

The use of foams has been studied in the past for decreasing gas mobility. Renewed interest exists in using foams to reduce carbon dioxide mobility. Foam stability and adsorption are areas that require additional research and field testing.

To assist in thermal recovery of the heavier crude oils, blocking agents such as foams may help to divert steam from steam channels to unswept areas. Surfactant alkaline chemicals and miscible or other gases may also assist in the thermal flooding process by improving displacement efficiency in unswept zones.

The more costly chemical floods require detailed reservoir knowledge to insure against the loss of expensive recovery agents. Also, the amount of chemicals needed could increase to the point of requiring a major manufacturing commitment from the chemical industry. It has been estimated that to achieve projected production rates in the United States, the demand for enhanced oil recovery chemicals will grow by as much as 34 percent per year to 1991. Specialty and commodity chemical demand growth would then be expected to be greater for enhanced oil recovery chemicals than for any other chemical market.[32] The domestic demand for diverting agents in steam flooding could reach several hundred million pounds by 1991, and over 100 million pounds of polymer chemicals are expected to be utilized. Alkaline chemical and micellar chemical consumption may each be in the 3-billion-pound range by 1991. It has further been estimated that the requirements for carbon dioxide and nitrogen will reach 1.4 trillion cubic feet by 1991.[33]

In carbon dioxide flooding, the large amount of gas needed for a major project is forcing many producers to explore and drill for carbon dioxide supplies. Carbon dioxide occurs naturally in underground reservoirs in the western United States and in the Gulf Coast region. In the west, it is formed by the intrusion of hot igneous rocks into carbonate sediments. The carbonate rocks are decomposed, forming carbon dioxide, which migrates and accumulates like natural gas. Recently, a pipeline was proposed to carry Colorado carbon dioxide to a Texas oil field to be used for an enhanced recovery project.

In enhanced recovery projects, the sharing of technology is a long-standing tradition. The free exchange of process development and field testing information avoids duplication and the waste of scarce engineering and scientific manpower and allows for the earliest possible application of technology to commercial projects. At the present time enhanced oil recovery technology is still not efficient or reliable and additional research is needed to realize the recoveries projected and expected in the next several decades.

Environmental Constraints

Air pollution from thermal recovery projects constitutes the most severe environmental impact of all enhanced recovery methods and is the single greatest constraint to the expansion of heavy oil recovery projects in the United States. The impact of thermal recovery methods on air quality varies depending upon the technology used, but all of the methods that burn oil to produce steam for stimulation or flooding, including in-situ combustion, produce pollutants. Each of these pollutants is regulated in the United States by the Environmental Protection Agency and in California, where most domestic heavy oil projects are located, by the California Air Resources Board and the appropriate county air pollution control district. The National Ambient Air Quality Standards to protect human health and to prevent significant air quality deterioration are, in most cases, less strict than California state or county requirements.[34]

The major pollutants discharged by thermal recovery projects are: sulfur dioxide, resulting from the combustion of oil in the steam generators; nitrogen oxides, resulting from the combustion of oil in the steam generators; hydrocarbons, emitted from the combustion of oil in the steam generators and with the steam from the wells upon release to the atmosphere; and particulates, emitted by the steam generators.

The regulations to control such air pollutants require costly new equipment such as scrubbing devices and vapor collection systems, the retrofitting of steam generators, and hard-to-get emissions trade-offs. There is the possibility that such requirements, along with disposal problems of the sludge produced by the scrubbing devices, will hamper the growth of heavy oil production, especially in California.

Enhanced recovery processes often rely on large quantities of water for steaming and flooding. For most processes the water must be fresh, but saline water can sometimes be used for flooding purposes. The consumption of fresh water will compete with domestic, agricultural, and industrial uses and also could result in serious depletion of surface waters in drier regions. The availability of fresh water for enhanced oil recovery processes may present problems in California, Texas, and western Louisiana where water supplies are limited.

The impact of chemicals on subsurface waters is also of environmental concern. Some chemicals used in enhanced oil recovery are considered toxic; thus, it is necessary to insure that they will remain in the reservoir and not contaminate groundwater supplies.

Enhanced Oil Recovery Economics

Estimates of the total domestic oil that may be recovered by the use of enhanced recovery methods are usually derived by applying a recovery factor (percentage) to the entire resource base. If these processes are not used in fields where applicable before these fields become abandoned, it is unlikely, due to increased costs, that new wells would be drilled and an enhanced recovery project started at a later time. The rate of abandonment of domestic target resources for enhanced oil recovery operations (in-place oil) has been estimated to be 10 billion barrels per year for the next 20 years. By the year 2000 this could result in one-half to two-thirds of the present domestic target resource for enhanced oil recovery having been abandoned.[35]

It has been estimated recently, in a report prepared for the U.S. Department of Energy, that steam drives fueled by lease crude oil produced on the site and polymer flooding conducted as part of an existing waterflood are the least expensive domestic enhanced recovery techniques. The report estimated that the cost of steam drives fueled by lease crude was between $21 and $28 per recovered barrel of oil and of polymers added to an existing waterflood between $22 and $28 per recovered barrel. Steam drives with purchased fuel were estimated to cost between $27 and $35 per barrel, in-situ combustion between $25 and $36 per barrel, carbon dioxide flooding between $26 and $39 per barrel, surfactant-polymer flooding between $35 and $46 per barrel, and polymer flooding as a tertiary technique (no existing secondary waterflood) $30 to $46 per barrel.[36] Cost estimates in the report are in mid-1980 dollars and apply to reservoirs ranging from the most favorable for the specific technique to those in which the technique would not be economical under current conditions. An oil price of $30 per barrel along with a 15 percent rate of return was assumed, which included considerations of such cost components as the windfall profits tax, royalties, and severance taxes.

In the United States there are currently about 226 sizable, active enhanced oil recovery field tests, 200 of which are funded by industry. The remaining 26 are cost-shared projects with the Department of Energy. There were an additional 423 incentive tests that received increased prices for produced oil over the controlled price. These tests were certified, however, before March 31, 1981, when all oil prices were decontrolled. It appears now that about 500 sizable, active enhanced oil recovery tests are in progress in the United States, when both industry and incentive tests are considered.[37]

Summary

The fraction of the in-place oil in an oil field that will flow unaided or can be pumped from the reservoir rock to the surface varies from as high as 80 to as low as 5 percent. The factors that determine the efficiency of such primary recovery are the energy (pressure) contained by the fluids in the reservoir, the viscosity (resistance to flow) of the oil, and the permeability (ease of flow permitted) of the reservoir rock. In the United States, primary recovery averages about 28 percent.

Secondary recovery methods attempt to maintain or restore reservoir pressure by the injection of gas or water (waterflooding). Depending upon reservoir conditions and oil properties, secondary recovery methods can improve overall recovery to between 30 and 50 percent of the in-place oil.

After secondary recovery methods have been used, the extraction of additional in-place oil from a reservoir will require the application of more sophisticated and expensive techniques. These techniques usually attempt to reduce oil viscosity and capillarity by the introduction of heat (in the form of steam) or of other injected substances such as carbon dioxide, nitrogen, polymers, solvents, surfactants, micellar fluids, and even bacteria in various combinations depending upon reservoir conditions and in-place oil properties. Attempts are also being made to mine the oil from depleted reservoirs. Enhanced oil recovery processes, sometimes called tertiary methods, can increase overall recovery to 40 to as much as 80 percent of the in-place reservoir oil, depending upon the process used and the physical properties of the reservoir and the oil. Enhanced oil recovery methods can be divided into five general types: thermal (steam soaking, steam flooding, in-situ combustion, and electric heating), chemical (polymer waterflooding, surfactant systems, and caustic flooding), gas (carbon dioxide, hydrocarbon miscible, nitrogen, and other inert gases), microbiological; and mining.

In 1980, enhanced oil recovery methods accounted for about 5 percent of total domestic oil production. More than three-quarters of this oil was produced by the injection of steam into heavy oil reservoirs. In the past ten years the number of domestic steam injection projects has more than doubled. The number of carbon dioxide injection projects also increased significantly during the 1970s. The general trend appears to be away from some of the more expensive chemical injection methods because of their high cost and some discouraging field results.

It appears that extensive use of enhanced oil recovery techniques could eventually double recovery from existing domestic crude oil reservoirs. Such recovery, however, would extend over many years so that in any one year enhanced recovery methods would probably continue to provide a relatively small (but increasing) fraction of total production. Thus, although this contribution could be significant, enhanced oil recovery cannot be expected to supplant conventional production, at least during this century.

At the present time enhanced oil recovery methods are still not efficient or reliable and additional research is needed to realize the expected recoveries. In addition, the amount of chemicals needed could increase to the point of requiring a major manufacturing commitment from the chemical industry.

Air pollution from steam injection projects constitutes the most severe environmental problem of all enhanced recovery methods and is the single greatest constraint to the expansion of domestic heavy oil recovery projects. The availability of fresh water for enhanced recovery processes may present a problem in areas where water supplies are limited, and the impact of enhanced recovery chemicals on subsurface waters is also of environmental concern. Since some of the recovery chemicals are considered toxic, it is necessary to insure that they will not migrate beyond the oil reservoir.

If enhanced recovery processes are not applied to old oil fields before they are abandoned, it is unlikely because of increased costs that new wells will be drilled and an enhanced recovery project started at a later date. By the year 2000 as much as one-half to two-thirds of the present domestic in-place oil resource targets for enhanced recovery may have been abandoned.

Notes

1. G. J. Demaison, "Tar Sands and Supergiant Oil Fields," *American Association of Petroleum Geologists Bulletin* (November 1977), pp. 1950–1952.

2. Joseph P. Riva, Jr., *Secondary and Tertiary Recovery of Oil* (Report prepared for the Subcommittee on Energy, House Committee on Science and Astronautics [Committee Print 40-382], U.S. Congress, Washington, D.C., October 1974), p. 24.

3. Todd M. Doscher, "Enhanced Recovery of Crude Oil," *American Scientist* (March-April 1981), p. 195

4. *Ibid.*

5. " 'Hot Plate' Heavy Oil Process Demonstration Due," *Oil and Gas*

Journal (July 23, 1979), p. 35; and "Field Work Begins on Unique Heavy Oil Process," *Oil and Gas Journal* (April 7, 1980), p. 40.

6. Aniekan Willie Iyoho, "Selecting Enhanced Oil Recovery Processes," *World Oil* (November 1978), pp. 62–63.

7. A. Herbert Harvey and Samy A. El-Feky, "Selective Reservoir Heating Could Boost Oil Recovery," *Oil and Gas Journal* (November 13, 1978), pp. 187–190.

8. Bennie Sinclair and William K. Ott, "Polymer Reduces Channeling, Ups Waterflood Recovery," *World Oil* (December 1978), pp. 102–104.

9. Doscher, "Enhanced Recovery of Crude Oil," p. 197

10. *Ibid.,* p. 198.

11. Shannon L. Matheny, Jr., "EOR Methods Help Ultimate Recovery," *Oil and Gas Journal* (March 31, 1980), p. 94.

12. Office of Technology Assessment, *Enhanced Oil Recovery Potential in the United States* (Washington, D.C.: Government Printing Office, 1978), p. 31.

13. National Petroleum Council, *Enhanced Oil Recovery* (Washington, D.C., 1976), p. 17.

14. "Caustics: The Inexpensive EOR," *Enhanced Recovery Week* (September 15, 1980).

15. National Petroleum Council, *Enhanced Oil Recovery,* pp. 13–16.

16. Doscher, "Enhanced Recovery of Crude Oil," p. 198.

17. *Ibid.,* p. 199.

18. R. E. Meissner, III, "Purifying CO_2 for Use in Enhanced Oil Recovery," *World Oil* (October 1980), pp. 189–198.

19. Matheny, Jr., "EOR Methods," pp. 114–116.

20. Keith Wilson, "Enhanced-Recovery Inert Gas Processes Compared," *Oil and Gas Journal* (July 31, 1978), pp. 162–166.

21. V. F. Coty, "Microorganisms of Oil Recovery: Status of Microbial Oil Recovery," *Proceedings of the 1976 Engineering Foundation Conferences: The Role of Microorganisms in the Recovery of Oil, November 9–14, 1975.* (Washington, D.C.: National Science Foundation, RANN), p. 77.

22. "Microbial Enhanced Recovery: Enhanced Oil Recovery and Improved Drilling Technology," Progress Review No. 25, DOE/BETC-81/1 (Washington, D.C.: Department of Energy, 1981), pp. 135–140.

23. *Ibid.*

24. *Ibid.*

25. Office of Technology Assessment, *Enhanced Oil Recovery Potential,* p. 33.

26. Mark Kindley, "Oil Mining—New Interest In An Oil Idea," *American Association of Petroleum Geologists Explorer* (December 1981), p. 30.

27. *Ibid.*

28. *Ibid.*

29. "DOE Report Examines Enhanced Oil Recovery Constraints," *Oil and Gas Journal* (April 28, 1980), p. 105.

30. *Ibid.,* p. 106.

31. *Ibid.,* p. 108.

32. "Big EOR Production Hike Seen From Chemicals," *Oil and Gas Journal* (August 17, 1981), p. 74.

33. *Ibid.,* pp. 74–75.

34. Elizabeth Martin and Randall Moory, *1985 California Oil Transportation Study* (Sacramento, Calif.: State Lands Commission, 1981), p. 26.

35. "Enhanced Oil Recovery Push Urged," *Oil and Gas Journal* (February 9, 1981), p. 34.

36. "DOE Updates Enhanced Recovery Economics," *Oil and Gas Journal* (July 13, 1981), p. 43.

37. Charles W. Perry, "The Economics of Enhanced Oil Recovery and its Position Relative to Synfuels," *Journal of Petroleum Technology* (November 1981), pp. 2033–2041.

4
Enhanced Recovery
of Unconventional Gas

All types of unconventional gas will be much less efficient to recover than conventional gas and will have to be produced at slower recovery rates. Unconventional gas is usually believed to occur in four geological environments: in tight (relatively impermeable) sandstones; in joints and fractures or absorbed into the matrix of Devonian-age shales; dissolved or entrained in hot geopressured waters; and in coal seams.

Tight-Gas Reservoirs

Tight gas is natural gas in either blanket or lenticular sandstones that have in-situ effective permeability of less than 1 millidarcy.[1] Historically, most of these relatively impermeable sandstones have been reservoir rocks for gas that are uneconomical due to low natural flow rates. Gas is widespread in many tight formations, but is commercial only in the better-quality reservoir rock. The field boundaries depend upon permeability and economics as well as upon structure and stratigraphy.[2]

Recently, the outlook for increased production of gas from tight sands has been enhanced by increases in gas prices and the use of massive hydraulic fracturing techniques, which create large collection areas in the low-permeability formations from which the gas can flow to a producing well. A fractured well, by exposing considerably more of a formation's rock face, can produce much more gas than an unfractured well. A fractured well in a tight-gas formation usually produces at a lower rate than a conventional gas well, but for a longer time.

Massive hydraulic fracturing is a volume designation only. Since tight-gas sands are massive, it is not possible to penetrate them effectively with the 30,000 to 50,000 gallons of fluid normally used in conventional reservoir fracturing. However, stimulation effectiveness can be improved with extremely large amounts of fluid and proppants (granular material, usually sand, used to keep the fractures open after the fluid has been removed) and with fluid-diversion techniques. Massive fracture stimulation can require a half-million gallons of gelled fluid and a million pounds of proppant sand. It is desirable to achieve a single propped-open vertical or near-vertical fracture that extends outward from 1,000 to 2,000 feet in opposite directions from the well. The height of the fracture can vary from 100 to 500 feet. Such fractures provide a pressure sink and channel the gas from the exposed rock face to the well.[3]

Exploration for tight-gas sands is expensive because reservoirs are not necessarily structurally controlled. Thus, seismic surveys often contribute little to the exploration effort. Pattern drilling must be used to locate good-quality reservoir rock. Better-quality reservoir rock is also sought by drilling along concentrations of natural fractures and by following lenticular reservoir trends and the trends of previous discoveries. Most drilling has been in blanket sand areas where the reservoir distribution is more predictable and the lateral continuity of the gas deposit improves recovery.

Twenty sedimentary basins that contain gas-bearing tight sandstones are located in North America, stretching from New Mexico northward into Canada and eastward into Arkansas and Louisiana. The locations of these tight-gas sand deposits have been known for over thirty years, but only the most favorable accumulations have been produced. The three basins with the greatest tight-gas sand exploration effort are the Piceance, the Green River, and the Uinta basins (see Figure 7.2). These basins, all in the Rocky Mountain area, are commonly referred to as the western tight-gas basins.

The major requirement for the commercialization of a large portion of the tight-gas sand resource is favorable economics. In addition, tight-gas sands have characteristics other than low permeability that limit commercial production. For example, they are often highly discontinuous or lenticular.[4] As formation permeability drops, recovery efficiency becomes highly sensitive to even very small permeability changes. When low permeability combines with lenticular formations, an additional problem arises. The principle commercial successes in tight formations have been in the relatively continuous blanket-type sands, which domestically produce about 85 percent of the tight-sand gas.[5] Most of the tight-gas sands, however, are notably discontinuous. Massive hydraulic fracturing has been successful in the discontinuous lenticular formations only where the individual lenses are large relative to the normal drainage area or are developed in conjunction with vertically adjacent blanket formations.[6] The section from which gas can be extracted may be dispersed in relatively thin (10 feet or less) zones in formations that range in thickness from 2,000 to 5,000 feet. The total producible section in such thick formations may be only a few hundred feet, with clays and shales often interbedded with the gas zones. Also, the gas-sand zones frequently contain a high percentage of formation water that impedes gas flow in the fracture systems. As porosities are low (usually between 5 and 15 percent), the relatively high formation water saturations of 40 to 70 percent reduce the gas-filled porosity in the sands from less than 3 to seldom over 9 percent.[7]

Finally, the gas zones in the tight formations often contain clays that swell when contacted by drilling or fracturing fluids, unless the fluids contain chemicals designed to inhibit such swelling. Existing methods and tools used for characterizing gas-producing formations have been found inadequate for tight-gas sands. The low permeability renders bottom-hole information from conventional drill-stem tests nearly useless for production estimates, and well logging has failed to distinguish gas zones from water zones.[8] Conventional laboratory core tests have vastly overstated actual permeability under reservoir conditions of water saturation and pressure. Present reservoir characterization is inadequate to design stimulation technology and to predict results.

The limited ability to determine fracture geometry constrains the control of fracturing performance. No techniques exist to measure fracture width or length. Improving the economics of tight-gas sand recovery depends upon increasing the reliability and efficiency of massive hydraulic fracturing. With advanced technology, it is projected that effectively propped fracture lengths may reach 4,000 feet, resulting in more efficient fracturing. Also, there is the expectation that the direction of the fractures can be determined in advance so that the wells can be located to maximize reservoir drainage. To further utilize this difficult resource will require that fracture technology also move toward a method of stimulation of all gas-zone intervals exposed to the well bore by creating multiple fractures engineered from the same well. Such fractures should intersect, in lenticular formations, sand lenses not initially in contact with the well bore. Thus geology and reservoir characteristics impose limits on the amount of gas that can be commercially recovered from the tight-sand reservoirs.[9]

Devonian Shale

The term Devonian refers to the geological time of deposition, and the rock unit, Devonian shale, is the general name for the shale strata which, in the United States, lie between the younger Berea sandstone and the older Devonian carbonates.[10] The Devonian shales were deposited about 350 million years ago in a shallow sea that covered almost half of the present continental land mass of the United States. The erosion of the lands adjacent to the sea produced massive quantities of fine sediments and organic debris that came to rest on the seafloor. Where the sites of deposition were in stagnant water (a reducing environment), the organic matter was preserved as a black organic-rich mud. Subsequently deposited overlying sediments combined with the natural heat flow of the earth resulted in sufficient pressure and temperature to gradually transform the organic mud into the black shales that are present today.[11] The Devonian shales were later uplifted and subjected to erosion and now cover only about one-fourth of the North American continent, where they occur mainly in the Appalachian, Michigan, and Illinois basins (see Figure 7.2). The chemical reactions caused by the heat and pressure that transformed the mud to shale also produced natural gas from the contained organic matter. Some of this gas migrated to and was trapped in adjacent sandstones to form conventional gas reservoirs; the rest remained locked in the nonporous shale.

The black and dark brown Devonian shale is rich in organic matter and has a much higher gas content than the poorer-organic gray Devonian shale. Thus, although almost all Devonian shale contains some gas, its gas-producing ability is generally indicated by color. The black and very dark brown shales appear to be the best gas producers, and most commercial Devonian shale gas production is derived from the brown shales. The production history of Devonian shale gas reservoirs indicates that the recovered gas occurs in well-connected fracture porosity. Recovery efficiency is quite high, 45 to 65 percent after thirty years of production. Production is generally at low flow rates but is long lasting.

With current techniques it is difficult to define both the areas of prospective Devonian shale and the zones within the shale that have the best gas production potential. The factor of greatest importance for commercial rates of production is the presence of natural fracturing. The only method currently available to detect areas that contain natural fractures is drilling. At present there is limited domestic exploration for Devonian shale; the majority of wells have been drilled in or near areas of known production. Normally, Devonian shale gas wells are drilled by conventional methods and are stimulated (fractured) by "shooting" with solid explosives. Recently, however, attention has shifted toward hydraulic stimulation and liquid explosives. Such treatment has met with varying degrees of success. In general, however, the estimated recovery efficiency is higher for the more expensive hydraulic fracturing (55 to 65 percent) than for shot wells (45 to 60 percent).

The major uncertainties concerning the future development of Devonian shale gas are mostly technical in nature. Exploration procedures for locating naturally fractured shale are not reliable, conventional production stimulation techniques are not always effective, and logging techniques often give conflicting results when used to attempt to locate potential producing zones.

To achieve higher production rates and increased ultimate gas recovery from Devonian shales, a research and development effort will be needed to improve the understanding of the locations in a basin where the Devonian shale is intensely fractured and thus a good gas-producing prospect. Also, the development and application of dual well completion techniques to produce marginal shales along with other gas-bearing sands would be significant. Still another problem that requires research concerns increasing the vertical efficiency of Devonian shale well stimulation techniques.

An additional constraint in some areas is a lack of pipelines. There are also many areas where only a portion of the total Devonian shale potentially available for gas production can be considered drillable. The domestic drillable acreage available on a county basis varies from about 30 to 90 percent, with an overall average of about 56 percent.[12] Factors contributing to nondrillable acreage include: physical barriers such as urban centers, lakes, and rivers; gas storage fields and other government-restricted lands; and lands with unavailable leaseholds.

Geopressured Brines

Deep geologically young sedimentary basins are usually undercompacted below depths of 6,000 to 10,000 feet. As sediments are buried, overburden weight increases, causing compression. The fluids contained in the pore spaces of such compressed sediments tend to be squeezed out. If the squeezed fluids cannot escape upward because the formation is sealed by overlying impermeable layers, the fluids will remain and bear part of the overburden load. The fluid pressure can then become quite high, sometimes almost double the normal hydrostatic gradient.[13]

The geopressured fluids of the Gulf Coast region of the United States occur in a sedimentary basin where the normal heat flow to the surface is impeded by insulating layers of impermeable shales and clays. The pressured fluids trapped beneath these layers become hotter than normally pressured formation fluids because upward fluid migration and heat flow have essentially been prevented for long periods of time. A portion of the heat that is continually rising from the interior of the earth is absorbed by the isolated fluids in the geopressured zones, causing them to become much hotter than is normal for their depth of occurrence. The heat and the pressure are potentially usable forms of energy.

Perhaps the most intriguing aspect of geopressured fluids is that most have been found to be saturated with 30 to 80 cubic feet per barrel of natural gas (primarily methane). The source of the natural gas is thought to be the marine shales within the geopressured zones. The solubility of methane in water increases with increasing pressure and temperature, but the presence of dissolved salts causes a marked decrease in methane solubility. The formation waters of the Gulf Coast are essentially brines. However, there is also likely to be a large amount of free but immobile gas in the geopressured zones. This combination of dissolved gas, which is limited by the salinity of the formation brines, and immobile gas is attracting interest to the geopressured deposits of the Gulf Coast (as most of the much richer conventional Gulf Coast gas fields have already been found).

Although most of the research on natural gas recovery from geopressured zones has occurred in the Gulf Coast, the search for oil and gas has led to the discovery of similar geopressured reservoirs in California, Utah, Colorado, Oklahoma, and Wyoming. Other geopressured formations have been found in Mexico, South America, the Far East, the Middle East, Africa, Europe, and the Soviet Union.

Drilling methods and equipment necessary to produce geopressured wells are considered adequate. To produce the gas, high flow rates of the hot geopressured brines must be maintained from areas of thick formations of high porosity and permeability and the erosion of equipment by sand present in the produced fluids must be controlled. However, the discovery of geopressured reservoirs large enough to commercially produce dissolved gas will be difficult. The geologic uncertainties concern fault barriers and reservoir sandstone continuity over the very large drainage areas required for economic development.[14]

There have been suggestions that it may not be necessary to produce enormous quantities of brine and then separate out the relatively small amounts of dissolved gas at the surface. There is an undemonstrated suggestion that, if the pressure in a geopressured reservoir is sufficiently lowered by fluid production, a point will be reached at which some of the dissolved gas will come out of solution and some of the existing free but immobile gas will also be released. The freed gas would then flow to the well bore faster than the formation water due to its higher mobility, with a resulting increase in the produced gas/water ratio. There is doubt, however, that the geopressured deposits contain sufficient amounts of gas to achieve this effect. In the absence of conclusive tests, the commercial production of free gas directly from a geopressured reservoir remains highly speculative. In fact, there is a possibility that the reduction of reservoir pressure will reduce permeability and thus adversely affect recovery.

The potential for technical improvements in geopressured gas recovery would parallel the experience of deep–gas-well drilling. It is uncertain if deeper drilling would enlarge the resource base, as the small amount of gas contained in geopressured brines appears to negate the potential benefits of technical improvements.[15] Mechanical problems may affect the longevity of production and hence the economics of a geopressured gas project. It is not known to what extent sand production may become a problem at such high-volume fluid production rates. Also, it is possible that serious down-hole corrosion and scaling problems could result from the presence of dissolved carbon dioxide. Because of the high rates of brine production required, treatment to prevent down-hole corrosion would be extremely difficult.

The two main environmental problems associated with the production of gas from geopressured deposits concern the proper disposal of the large volumes of formation brines necessary to derive enough gas for commercial production and the possible land subsidence that may occur upon the withdrawal of such large volumes of underground fluids. Slight subsidence, which would scarcely be noticed in many areas, could be serious in the low-lying regions of the Gulf Coast. The disposal of produced brine into the same formation from which it came could solve this problem, but the cost of deep injection wells would seriously reduce the economic attractiveness of geopressured development.[16]

Large-scale production of gas from geopressured brines is unlikely, at least in the near term, because of: extremely high capital investment and operating expense per unit of gas production; the small amount of dissolved gas in each barrel of brine, which will require that each well be capable of producing at high fluid rates for many years; the low recovery efficiency of about 3 percent of the in-place gas; and the highly faulted condition of the geopressured reservoirs, which greatly limits the size of the resource available for exploitation.[17]

Coal Seams

Considerable quantities of methane are trapped within eastern and western coal seams. Little use has been made of this coal gas domestically, but coal-bed gas production is common in Europe, although the gas is often mixed with air. The methane in coal seams was generated naturally during coal formation. Although much of the gas that formed during the initial coalification process is lost to the atmosphere, a significant portion remains as free gas in the joints and fractures of the coal seam, as adsorbed gas on the internal surfaces of the micropores within the coal itself, and as desorbed gas in adjacent strata, which may act as supplementary reservoirs.[18] Since coal is relatively impermeable, any methane that is recovered must, in general, flow through existing fracture systems in the coal. Therefore, coal seams that are highly fractured appear to be the best sources of coal-bed methane.

Some of the gas entrapped in and around coal seams can be liberated either by premining degasification or by the act of mining itself. Large amounts of gas are often liberated during the mining of the more gassy coal seams. Thus, the impetus for degasification has primarily been the result of safety considerations, although it has long been recognized that coal-seam gas could have value as an energy source if it could be commercially recovered and utilized. The gas liberated at the mine face where the coal is cut is carried away in the mine ventilation system in an extremely diluted form and probably is not economic to recover, even at higher gas prices. The gas that slowly bleeds from the remaining coal seams or wall rock after mining is completed is usually in a more concentrated form, but it sometimes contains large amounts of air and other contaminants. Effective premining gas drainage will reduce this bleeding gas, but there may be situations in which it could be used as a local energy source.[19]

The gas that can be obtained from premining draining of coal seams is considered to have greater energy potential. The major techniques for the recovery of this type of coal gas are vertical wells, horizontal holes from mine access, horizontal holes from mine-shaft bottoms, and slant holes. The technique of recovering coal-bed gas by means of vertical holes from the surface usually requires hydraulic fracturing to increase the limited gas flow, unless the seam is naturally highly fractured. However, the extension of hydraulically created fractures into the overlying rock can constitute a hazard when the coal is eventually mined, although there have been reports of at least ten hydraulically fractured seams that were later successfully mined.[20]

The process of drilling horizontal holes into virgin coal from the working sections of an underground coal mine is generally considered the most economical method of recovering coal-bed gas. The gas thus liberated is conveyed by a piping system to the surface where it can be collected.

Another technique involves sinking vertical shafts ranging from 8 to 20 feet in diameter to the bottom of the coal seam. Horizontal holes are then radially drilled from the shafts into the coal to intersect the natural fractures and collect the trapped gas. This concept is potentially viable but has a large initial investment in shaft sinking. As there is no fracturing, there would be no adverse impact on the safe minability of the coal seam.

The slant-hole technique appears to be the most difficult and costly. In this method a small-diameter vertical hole is drilled from the surface and intentionally deflected to penetrate the coal seam parallel to its bedding plane. The hole then continues in the coal horizontally. The risk of failing to properly deflect the hole to keep it within the coal seam for long distances is quite high. The dewatering of the slant holes is another problem that is not yet fully resolved. Improvements in drilling and dewatering techniques are necessary before this method can be considered commercial.

Coal-seam gas is normally vented during mining as a safety measure. The initial domestic target for recovering natural gas from coal seams is the 0.08 trillion cubic feet of methane emitted annually from working coal mines. Some of the domestic attempts at coal-bed gas recovery have been successful, but none has established a long-term economic viability as an energy project. There has been speculation that the Appalachian basin coal seams may be too thin and contain too little gas to support commercial recovery. A better target for recovering gas from coal may be the thick, bituminous coal beds of Colorado and other western states.[21]

The most important environmental problem associated with the production of coal-seam gas is the disposal of coal-bed water, which varies widely in composition from slightly acidic to slightly alkaline and is often saline. Environmental regulations vary, but if large quantities of such water had to be removed and treated before disposal, substantial costs would be involved. The availability of water may become an issue in the western United States where gas recovery projects could cause groundwater depletion problems.

Much of the recovery technology for coal-seam gas is proved, although the drilling guidance systems for slant holes need improvement, particularly the continuous in-hole surveying instrumentation. Production risk is not in the equipment but rather in the inability to estimate gas flow rates. There is also a basic lack of baseline information and of basic research on the gas content of different types of coal and on the factors that control the flow of gas in coal beds.[22] A source of conflict between the mining of coal and the production of coal-seam gas is the danger of the coal seam being rendered economically unminable by fractures created hydraulically for the purpose of increasing gas production. To solve this problem, the evaluation of potential commercial production of gas from coal seams could include ways to enhance mining safety and productivity. In spite of its widespread occurrence, however, large-scale gas production from domestic coal seams is not likely in the near future.

Summary

Current annual domestic production of unconventional gas is about 1 trillion cubic feet from tight gas reservoirs and about a tenth of that amount from Devonian shale. There is no current commercial gas production known to be derived from a geopressured deposit. Although some domestic coal-seam gas recovery projects were successful, none has established a long-term economic viability as an energy project. Coal-bed gas recovery results have been somewhat unpredictable, and domestic coal gas recovery is currently negligible. Thus, about 5 percent of domestic natural gas production is currently derived from unconventional sources.

Although the estimates of undiscovered recoverable resources of domestic unconventional gas are the same as undiscovered recoverable resource estimates of conventional domestic gas, only a very small portion of the former have actually been proved as reserves. The estimated in-place gas volumes are considerably greater than the undiscovered recoverable resource volumes for conventional gas, as a smaller portion of the total gas can be extracted from such deposits than from a conventional gas reservoir.

In the longer term there may be sufficient unconventional domestic gas resources to last for many decades, but this optimistic projection is based upon very large recoveries from geopressured brines, the most difficult of the unconventional gas resources to produce and therefore to estimate, and from gray Devonian shales. There may well be an even smaller undiscovered recoverable unconventional domestic gas resource than an undiscovered recoverable conventional domestic gas resource, which would be about thirty years of an unconventional gas supply at current production rates. The unconventional gas sources will be much more expensive to exploit and will have to be produced at much slower rates than conventional gas fields.

A prudent view of unconventional domestic gas is that it is a supplemental supply rather than a vast resource that will supplant declining conventional gas production. Gas production of such magnitude from domestic unconventional sources and the existence of sufficient reserves to sustain such production have not yet been demonstrated. Unconventional gas deposits will have extremely variable physical properties and recoveries will be low. Each deposit will have to be treated individually. Considerably more time and effort will be needed to produce a volume of unconventional gas compared to an equal volume of conventional gas, and the unconventional gas production will have to be spread over a larger time period to accommodate the much slower delivery rates of this resource.

Notes

1. A millidarcy is a unit of permeability, a measure of the ease with which fluids can flow through porous rock.

2. National Petroleum Council, *Unconventional Gas Sources: Tight Gas Reservoirs,* Part I (Washington, D.C., 1980), p. 5.

3. Board on Mineral Resources, Commission on Natural Resources, National Academy of Sciences, *Natural Gas from Unconventional Geologic Sources,* Report FE-2271-1 (Washington, D.C.: Energy Research and Development Administration, 1976), pp. 128–129.

4. Velo A. Kuuskraa, J. P. Brashear, Todd M. Doscher, and Lloyd Elkins, "Vast Potential Held by Four 'Unconventional' Gas Sources," *Oil and Gas Journal* (June 12, 1978), p. 48.

5. *Help For Declining Natural Gas Production Seen in the Unconventional Sources of Natural Gas,* Report EMD-80-8 (Washington, D.C.: General Accounting Office, 1980), p. ii.

6. Kuuskraa et al., "Vast Potential," p. 49.

7. *Ibid.*

8. National Petroleum Council, *Unconventional Gas Sources: Executive Summary* (Washington, D.C., 1980), p. 30.

9. Joseph P. Riva, Jr., *Unconventional Gas; Pursuing Energy Supply Options: Cost Effective R&D Strategies* (Report prepared for the Joint Economic Committee [Committee Print 71-9900], U.S. Congress, Washington, D.C., April 27, 1981), p. 191.

10. National Petroleum Council, *Unconventional Gas Sources: Devonian Shale* (Washington, D.C., 1980), p. 13.

11. *Ibid.*

12. *Ibid,* p. 73.

13. William J. Bernard, "Deep Geopressured Aquifers: A New Energy Source?" *Petroleum Engineer International* (March 1978), p. 86.

14. National Petroleum Council, *Unconventional Gas Sources: Geopressured Brines* (Washington, D.C., 1980), p. 21.

15. *Ibid,* p. 14.

16. Riva, *Unconventional Gas,* p. 191–192.

17. National Petroleum Council, *Geopressured Brines,* p. 4.

18. National Petroleum Council, *Unconventional Gas Sources: Coal Seams* (Washington, D.C., 1980), p. 5.

19. *Ibid,* p. 7.

20. *Ibid.,* pp. 15–16.

21. Kuuskraa et al., "Vast Potential," pp. 51–52.

22. National Petroleum Council, *Coal Seams,* pp. 43–44.

5
Reserves, Resources, and Reserves/Production Ratios

When a new reservoir is discovered it is obviously important to determine how much oil and/or gas it contains in order to justify undertaking the expense of development and production. Natural petroleum accumulations occur in a wide variety of sizes from a few super-giants with over 5 billions barrels of oil-equivalent to numerous tiny fields with as little as a few thousand barrels. Naturally, the minimum size of a reservoir worth developing depends upon a variety of factors such as drilling and production costs, proximity to markets, and petroleum prices. There is, however, a broader consideration.

The importance of petroleum as a world energy source is difficult to overdramatize: it provides 62 percent of the world's energy supply. The growth in energy production during the twentieth century is unprecedented and increasing oil production has been, by far, the major contributor to that growth. At the same time, until very recently, the declining price of oil in real terms has helped to stimulate a parallel growth in economic activity. An immense and intricate system moves 59 million barrels of oil from producer to consumer every day. The production and consumption of the world's oil is of vital importance to international relations and has frequently been a decisive factor in the determination of foreign policy. A country's position in this system is determined by its production capacity as related to its consumption. Production prospects depend upon reserves, so it is important for a country to have an inventory of its petroleum resources. The possession of significant oil deposits is sometimes the determining factor between a rich and a poor country, and for any country the presence or absence of oil deposits is of major economic consequence.

Proved Reserves

Proved oil reserves are those quantities of in-place oil (measured in terms of stock tank barrels of 42 U.S. gallons at atmospheric pressure—14.73 pounds per square inch—and corrected to 60 degrees Fahrenheit) that have been identified and are considered, on the basis of geological and engineering knowledge, to be recoverable under current economic conditions with existing technology. The producibility should be supported by either actual production or conclusive formation tests. The area of an oil reservoir that is considered proved must be delineated by drilling and defined by gas-oil or oil-water contacts. The immediately adjoining portions of the reservoir that are not yet drilled, but which can reasonably be judged as producible on the basis of available geological and engineering data, may also be considered proved. In the absence of information on fluid contacts, the lowest known structural occurrence of oil is considered the lower proved limit of the reservoir.

Proved natural gas liquids reserves are defined in a similar manner. They are the current estimated quantity of natural gas liquids that analysis of geologic and engineering data demonstrates with reasonable certainty to be recoverable in the future from known gas reservoirs under existing economic and operating conditions. Natural gas liquids are liquid hydrocarbons in natural gas deposits that are separated from the gas either in the reservoir, through retrograde condensation, or at the surface in field separators or gas processing plants through condensation, absorption, adsorption, or other modification. The liquids usually consist of propane and other heavier hydrocarbons and are commonly referred to as condensate, natural gasoline, or liquefied petroleum gas.

Proved natural gas reserves are the total volume of natural gas estimated to be recoverable from known reservoirs under the economic and operating conditions existing at the time of the estimate. Such volumes of gas are expressed in cubic feet at an absolute pressure of 14.73 pounds per square inch and a temperature of 60 degrees Fahrenheit. Reservoirs are considered proved that have demonstrated the ability to produce by either actual production or by conclusive formation test. The area of a reservoir considered proved is that portion delineated by drilling, including adjoining portions known through geological and engineering assessment. It is important to note that improved recovery technology and/or higher petroleum prices will increase existing petroleum reserves.

In the period immediately following a petroleum discovery, and in the early days of development, only the volumetric method of reserve evaluation is available.[1] The first step is to define the limits of the petroleum-bearing structure and obtain its total volume. In some areas, however, a large total amount of petroleum may be distributed among a number of small independent reservoirs. It then becomes a matter of some difficulty to define the limits of each without drilling an excessively large number of wells, many of which may be poorly located for production wells or may even be dry holes. Next, it is necessary to determine the average porosity of the reservoir rock as well as reservoir pressure, temperature, gas/oil ratio, and water saturation. From this information, the amount of reservoir in-place gas and/or oil can be calculated and converted to proved reserve volumes at atmospheric pressure and temperature.

There is considerable scope for error in a calculation of this type, especially when it is based on limited data. The geological and geophysical evidence may be inadequate and the formation characteristics determined from the logs and cores, especially if from only a few wells, may not be representative of overall reservoir conditions. As a field is developed and becomes further defined revision of earlier proved reserves estimates is often necessary.

After a field has been in production for some time, it may be possible to check the estimate of in-place petroleum by means of a "material balance" calculation. If a closed tank containing known amounts of oil, gas, and water is considered, it is possible to calculate the decline in pressure caused by the removal of known amounts of each fluid. Conversely, if the change in pressure is measured after various known increments of production, the original amounts of oil, gas, and water in place in the reservoir can be calculated. This method is very sensitive to, and subject to, a wide range of errors in the determination of average pressure. Such errors are frequent and can arise in a number of ways; thus, in practical application, the results of material balance calculations must be viewed accordingly.[2]

The petroleum in place is only one factor that must be estimated in any consideration of proved reserves. It is also necessary to estimate how much of the in-place petroleum can be recovered. In the case of oil, the percentage recoverable from a reservoir varies widely depending upon the pressure in the reservoir, the viscosity of the oil, and the porosity and permeability of the reservoir rock. Up to 80 percent of the in-place oil in a reservoir may flow unaided into a well; this initial production is known as primary recovery. Secondary and tertiary (enhanced) recovery methods are necessary to achieve a higher percentage recovery of in-place oil than is indicated by initial production rates.

Gas reservoirs, like oil reservoirs, differ greatly from one another in shape, size, formation characteristics, and fluid properties. All of these variations affect reservoir performance and recovery. In the case of a natural gas (single-phase) reservoir at moderate temperature and pressure, it should be possible to recover all of the in-place gas by dropping the pressure sufficiently. If the pressure is effectively maintained by the encroachment of formation water, however, some of the gas will be trapped by capillarity behind the advancing water front. Thus, no more than about 80 percent of the gas will be recovered in practice. On the other hand, if pressure declines, there is an economic limit at which the cost of compression exceeds the value of the produced gas. The production of natural gas is much more efficient than the production of oil. Depending upon formation permeability, recovery can be as high as 75 to 80 percent of the original in-place gas.

After a reservoir has been fully developed and under production for some time, it is often possible to obtain an estimate of the remaining producible reserves by extrapolating a suitable curve of past performance down to the economic limit of recovery. If secondary or tertiary recovery methods are applied, the reservoir will produce additional quantities of oil. A well or field will be abandoned when the cost of producing additional oil or gas exceeds its value, even though it is still technically possible to extract more petroleum.

Inferred Reserves (Field Growth)

Inferred reserves are those that should eventually be added to proved reserves through extensions of known fields, revisions of earlier reserve estimates resulting from new subsurface and production information, and production from new producing zones in known fields. Also included in inferred reserves are additional reserves in known productive reservoirs in existing fields expected to respond to improved recovery techniques, such as fluid injection. These reserves are sometimes known as indicated reserves.

Inferred reserves can be estimated by extrapolating the rate of growth of discovered petroleum volumes for the region in question, using correction factors based on the time lapse since the initial year of discovery. Estimates of oil and gas reserves in existing fields often grow with time. The early estimate of a field's potential is somewhat speculative compared to a later calculation of proved reserves made with additional data on strict reservoir engineering guidelines. Nevertheless, reevaluations of such original estimates after several years of development have supported their general validity. Accumulated domestic original reserve estimates average about 50 percent of the final estimates. There is an initial tendency to overestimate small discoveries and to underestimate large ones. A very few upward adjustments for large fields will overcompensate for the more common downward adjustments for many small fields. These adjustments are indicated in reserve figures each year as revisions. (Some recently discovered large fields, however, have not grown. The early reserve estimates for the super-giant Prudhoe Bay field in Alaska, for example, have remained essentially unchanged through several years of production.)

In foreign areas where even proved reserve estimates are approximate and fluctuating, inferred reserves can be estimated by assuming that the higher of the published reserve estimates include field growth and the lower estimates more nearly correspond to proved reserves. The wide variability of data used and the fact that proved reserves are also estimates cause a significant degree of uncertainty in the calculations of inferred reserves.

Undiscovered Recoverable Resources

The amount of petroleum ultimately recoverable from a sedimentary basin is determined by the volume originally generated from the indigenous organic matter and by the geologic history of the basin. The amount of petroleum theoretically formed may be estimated from the volume and quality of the source rock and the geothermal history of the basin, and then all available lithologic, tectonic, hydrodynamic and physical data can be employed to arrive at a projection of the quantities that may be expected to be trapped and ultimately recovered. However, undiscovered petroleum resource estimates necessarily have a low reliability. They often depend on geologic projections of such great uncertainty that an unexpected change in the thickness of a source bed or in the continuity of a reservoir horizon can affect their accuracy by a significant factor.

However imperfect, projections of undiscovered petroleum resources are necessary for an understanding of the evolution of petroleum development within a given basin and for the formulation of future energy policy. There are three basic types of petroleum resource estimation: the volumetric methods, the geologic play-analogy methods, and the past-performance extrapolation methods.

Volumetric methods have been used for many years. The sediment-volumetric method projects the amount of petroleum known to exist in a developed region to an unknown region of apparently similar geologic character. This method frequently leads to overly optimistic resource estimates, especially when it is assumed that undrilled rock volumes in little-known areas may contain the same amount of petroleum as an equal volume of rock that has already been drilled in a producing area. This assumption does not recognize that the potential of many undrilled areas has already been geologically discounted by the petroleum industry. These one-to-one comparisons have sometimes been reduced by one-half in more recent estimates, but even the one-half value may lead to inflated resource projections.

Another volumetric approach has been to multiply the area of a potentially productive basin by the estimated net reservoir thickness and by a projected yield in barrels per cubic mile. A problem with this method is that basins with large volumes of sediments can be almost totally lacking either in adequate reservoir rocks or in good source rocks. In either case, the actual amount of recoverable petroleum may be nearly zero, though a volumetric assessment with even a modest yield estimate could suggest the possibility of billions of barrels. Still, the volumetric method of basin assessment is useful in early exploration stages when data are scarce.[3]

The basic geologic play-analogy technique for estimating undiscovered petroleum resources is to compare the factors that control known occurrences with the factors present in prospective areas. In making evaluations of basins or sub-basins or smaller areas it is often useful to consider the resources in terms of actual or potential producing trends or plays and individual prospects. Trends are belts in which the conditions for oil and gas production are especially favorable. These conditions would include good porosity and permeability in reasonably thick reservoir rock; suitable traps; and favorable source rocks, geothermal gradients, and hydrodynamic conditions. Within such trends the oil and gas reservoirs will usually be of similar age and lithology and in similar types of traps and contain petroleum of related chemical nature.

A play is a practical planning unit around which an exploration program can be constructed. It has geographic and stratigraphic limits and is confined to a formation or group of closely related formations on the bais of lithology, depositional environment, or structural history. Play-analysis methods usually have been applied to relatively small areas but can be expanded to apply to entire geologic horizons or stratigraphic units for the appraisal of a total basin or province. Recently, play-analysis methods have been applied to frontier areas where geological and geophysical data are limited and the geological variables have been described by subjectively derived probability functions based on the judgment of a group of estimators through the use of selected analogs.[4]

The most recent U.S. Geological Survey method of assessing undiscovered resources includes the extrapolation of known producibility into untested sediments of similar geology. It also includes: volumetric techniques using geologic play analogs with the setting of upper and lower yield limits through comparisons with a number of known areas;

volumetric estimates with an arbitrary general yield factor, used when direct analogs are not known; a graded series of potential-area categories; and a comprehensive comparison of all published estimates for each area to all estimates generated by the above methods. Individual and collective appraisals are made for each region and Monte Carlo approximation techniques are applied to derive the probability function for the amount of undiscovered petroleum in each region assessed.[5]

Monte Carlo is a procedure that simulates probability distributions by running many trials. The range of possible answers reflects different combinations of values selected at random from within specified ranges of input parameters. The computer simulation multiplies many possible combinations, giving a probability curve. The Geological Survey's estimates for undiscovered recoverable petroleum resources are given as a range, the low value of which is the quantity associated with a 95 percent probability (a 19 in 20 chance) that there is at least this amount of petroleum. The high value is the quantity with a 5 percent probability (a 1 in 20 chance) that there is at least that amount. A statistical mean is mathematically derived from the Monte Carlo technique, and these amounts may be mathematically totaled. (The statistical mean is the expected amount of undiscovered petroleum, taking into account the very low probability of the discovery of a very large field that will greatly impact the region's reserves.)

The following geological parameters are emphasized in U.S. Geological Survey assessments as being essential for significant oil or gas accumulations.

1. An adequate thickness of sedimentary rocks.
2. The presence of source beds with considerable amounts of dispersed organic matter.
3. A suitable environment of the maturation of organic matter.
4. Porous and permeable reservoirs.
5. Hydrodynamic conditions that were favorable for early migration and ultimate entrapment.
6. Favorable geothermal history.
7. Adequate trapping mechanisms.
8. Favorable timing of petroleum generation and migration in relation to the development of a trap.

In regions where some of the essential geological parameters, trends, and plays cannot yet be assessed, special attention is given to oil and gas seeps, degree of variability of rock types, organic-rich beds as potential source rocks, progressive growth of structural features, unconformities, and evaporites.

The Potential Gas Committee, under the auspices of the Potential Gas Agency of the Colorado School of Mines Foundation, Inc., and the gas industry, prepares estimates of the potential natural gas supply for the United States. The estimates are made by the use of the geological analogy–volume of sediments technique, which is basically a comparison of the factors that control known occurrences of gas with the factors present in prospective areas. Known occurrences are expressed as the volume of gas that is ultimately recoverable per unit volume of reservoir rock within an adequately explored portion of a geological province. Such known relationships, with appropriate adjustments for variations in geological and reservoir conditions, are then attributed to incompletely explored sedimentary rocks in the same or similar geologic provinces.

Mobile Oil Company has employed a probabilistic engineering analysis for partially explored to well-explored plays. It includes an analysis of production and reserve data and of discovery index and exploration success. In less-explored regions geological parameters and analogs are used, as are sediment volumes. The method attempts, wherever possible, to use the play as the basic unit for evaluation. Computer input for Mobil's petroleum resource estimates uses a Delphi-like approach in which input estimates are challenged to bring out the basis of the estimates and to improve their quality. Many possible combinations of the various geological inputs are obtained and the results are aggregated by Monte Carlo procedures into a probability distribution of the petroleum potential of a given region.[6]

Past-performance or behavioristic extrapolation methods are based on the extrapolation of past experience with such indicators as discovery rates, cumulative production, or productive capacity curves. By the use of mathematical derivations, past performances are then fit into logistic curves that can be projected into the future. A decline in any of the chosen parameters could signal a declining resource base. In general, past performance extrapolation methods are most applicable in the later stages of petroleum exploration in mature areas such as the lower forty-eight states. They are less valid in frontier regions where little history exists or in areas that are not geological or economic replicas of the historical model.

The M. King Hubbert growth curve projections have proved relatively accurate in the forecasting of the crude oil production peak in the United States. In the mid-1950s the United States petroleum industry had been in operation for just under 100 years and cumulative oil production had amounted to about 52 billion barrels. If future production were to total three or four times past production, it was generally felt that production would continue its expansion for many decades. At that time, Hubbert, using his oil production curves with an estimate of 150 to 200 billion barrels for ultimate United States crude oil production, arrived at the conclusion that United States production would peak between 1966 and 1971 (depending upon whether the 150- or 200-billion-barrel figure was used). United States domestic oil production peaked in 1970, as Hubbert had predicted. Thus extrapolations have an important place in petroleum resource assessments and have the advantage of being directly tied to the realities of experience. Hubbert showed that, in the United States, the production curve lags behind the discovery curve by about 10–12 years.[7] Where the two curves cross, discovery additions and production subtractions are equal and reserves begin to decline; this decline is followed by a decline in production.

The extrapolation of discovery rates (in barrels discovered per foot of exploratory drilling) has been further refined in specific regions of the United States and then given broader application to the world as a whole.[8] Exxon has used historical yearly discovery rates per foot of new-field exploratory drilling in conjunction with field growth studies to project undiscovered petroleum resources. Such resource investigations also emphasize the geology of both plays and provinces and use Monte Carlo methods to aggregate speculative assessments of various plays into probability graphs for larger provinces and for the total United States.

Reserves/Production Ratios

The reserves/production ratio is a measure of the rate of petroleum production. The United States produces its oil at a faster rate than any other major producer nation, with more than 10 percent of domestic reserve stocks taken from the ground annually (an average reserves/production ratio of 9:1). The average world production rate is only about one-third of the U.S. rate (triple the domestic reserves/production ratio). Although individual fields vary in the rates at which they can be produced, an average reserves/production ratio of 10:1 is near the maximum annual oil withdrawal rate permitted by the physical properties of reservoirs without a loss of ultimate production. As oil fields near depletion, reserves/production ratios can drop considerably below 10:1, to as low as 5:1 or even 3:1.

Average reserves/production ratios much below 10:1 to 8:1 are an indication of an old producing region in which (in the absence of significant new discoveries) continued production declines may be expected. On the other hand, average reserves/production ratios of around 10:1 occur in areas of mature, intensive development in which oil is being produced as fast as physically possible without risking reservoir damage that would result in an ultimate recovery decline. Regions of less mature and intensive development usually have reserves/production ratios averaging around 15:1, although some producer countries with very large fields and low production (because of conservation practices) have average reserves/production ratios several times higher than 15:1.

Nonassociated natural gas can be produced at faster rates than oil. Average reserves/production ratios for intensively developed, mature regions may range between 5:1 and 10:1. In the United States the average reserves/production ratio for gas is currently 10:1. In most petroleum-producing areas of the world, gas production is not nearly as intensively developed as in the United States, and average gas reserves/production ratios are much higher than they are in the United States.

Summary

Proved petroleum reserves are those quantities of in-place petroleum that have been identified and are considered, on the basis of geologic and engineering knowledge, to be recoverable under current economic conditions with existing technology.

Inferred reserves are those that should eventually be added to proved reserves through extensions of known fields, revisions of earlier reserve estimates resulting from new information, and production from new producing zones in known fields. Inferred reserves can be estimated by extrapolating the rate of growth of discovered petroleum volumes using correction factors based upon the time lapse since the initial year of discovery.

The amount of petroleum ultimately recoverable from a basin is determined by the volume generated from the indigenous organic matter and by the geologic history of the basin. The three basic methods of estimating undiscovered recoverable petroleum resources are volumetric, geologic play analogy, and past-performance extrapolation. Newer assessments combine these methods and use computer techniques to produce probability ranges for the resource estimates.

Reserves/production ratios are a measure of the rate of petroleum production. Average oil reserves/production ratios much below 10:1 to 8:1 are an indication of an old producing region in which continued production declines may be expected. Average oil reserves/production ratios of around 10:1 occur in mature and intensively developed producing regions. Regions of less-intensive development usually have reserves/production ratios that average around 15:1, but some producer countries with very large fields and relatively low production (due to conservation or cartel practices) have average oil reserves/production ratios several times higher than 15:1. Nonassociated natural gas can be produced at faster rates than oil. Average gas reserves/production ratios for regions with intensive development may range between 5:1 and 10:1. In most of the world's petroleum-producing areas, however, gas production is not nearly this intensive and average gas reserves/production ratios are much higher.

The next five chapters will consider the petroleum prospects of the following five important petroleum-producing regions of the world: Indonesia (a typical history of a region with maturing oil production), the United States (a region with declining production and a vast amount of available data), the Soviet Union (a region about to begin an oil production decline), Mexico (a region with excellent production prospects), and the Arabian-Iranian basin (a region containing the world's greatest volume of oil). The geology, production history, reserves, and resources of these five regions in various stages of petroleum development will provide insights into the process of world petroleum exploitation and its relationship to the distribution of world petroleum resources.

Notes

1. G. D. Hobson, ed., *Modern Petroleum Technology* (New York: John Wiley & Sons, 1973), p. 155.

2. *Ibid.,* p. 156.

3. David A. White and Harry M. Gehman, "Methods of Estimating Oil and Gas Resources," *American Association of Petroleum Geologists Bulletin* (December 1979), p. 2187.

4. Betty M. Miller, "Application of Exploration Play-Analysis Techniques to Assessment of Conventional Petroleum Resources by the USGS," *Journal of Petroleum Technology* (January 1982), p. 55.

5. Betty M. Miller et al., *Geological Estimates of Undiscovered Recoverable Oil and Gas Resources in the United States,* Circular 725 (Reston, Va.: U.S. Geological Survey, 1975), pp. 20–25.

6. "World Crude Resource May Exceed 1,500 Billion Barrels," *World Oil* (September 1975), p. 56.

7. M. King Hubbert, "Degree of Advancement of Petroleum Exploration in United States," *American Association of Petroleum Geologists Bulletin* (November 1967), p. 2207.

8. David H. Root and Lawrence J. Drew, "The Pattern of Petroleum Discovery Rates," *American Scientist* (November-December 1979), pp. 648–652.

6
The Petroleum Prospects of Indonesia

Two overriding principles apply to the petroleum contained in sedimentary basins. First, most oil and gas are contained in a few large fields, but most fields are small. Second, as exploration progresses, the average size of the petroleum fields being found decreases significantly, as does the amount of petroleum found per unit of exploratory drilling. (The large fields are usually found first.) These principles are illustrated in the sedimentary basins of Indonesia.

History

The history of Indonesian petroleum exploration and development can be divided into several phases.[1] The pioneering phase began in the 1880s and lasted into the early 1930s. Such early work was confined to the onshore areas, mainly the larger islands. The second phase, the modern onshore phase, took place from about 1930 to 1965. The first phase of exploration was accomplished mostly by surface mapping, shallow-core drilling, and the drilling of oil seeps; the second phase was marked by the introduction of geophysical techniques, especially seismic surveys, permitting deeper prospect evaluation. The fields that were discovered early were generally rather shallow in depth, but in 5 of the 11 producing Tertiary basins, the largest field was discovered before 1930, in the preseismic phase.

The third phase of exploration began in 1966 with the signing of the first offshore production sharing contract. This new phase was made possible by advances in offshore drilling and production technology. Although most of the work from the mid-1960s to the mid-1970s was done offshore, extensive onshore exploration was also going on, usually in remote areas made accessible by the use of helicopters. Exploration was mostly done by seismic surveys, but geological mapping continued to play a necessary part. This phase of accelerated exploration activity, sometimes called the "quiet boom," came to an end in 1974 with a significant downward trend in oil exploration. Throughout the middle 1970s exploration activity was low due to a general reorganization of the Indonesian state oil company, Pertamina, which resulted in the renegotiation of contracts with the international oil companies to provide an increased government share. The latest phase began in 1978 with governmental assurances of fiscal stability and conciliatory incentives.[2] By mid-1980, Indonesia was experiencing an exploration boom. Recently, however, certain conversions from contracts to production-sharing agreements have stalled. Failure to reach an agreement considered fair by both sides could hurt Indonesia's reputation with the oil industry and may again have a negative impact on foreign investment.[3]

Geology

Zones of earthquake and volcanic activity can be used to divide Indonesia into four large regions that are parts of tectonic plates (see Figure 1.5). The Eurasian cratonic plate contains most of the island system: Sumatra, Java, Kalimantan, and a part of Timor and Sulawesi (the Celebes). The Indian-Australian plate contains the remainder of the island system, Timor and Irian Jaya, as well as the bounding Java trench. The Philippine Sea and the Caroline plates bound the Indonesian land areas to the northeast.

Superimposed upon these huge plates are 28 Tertiary basins (see Figure 6.1), mostly initiated by tectonic movements at the end of the Mesozoic. Little is known about the Mesozoic rock sequences that locally may underlie these basins. Of the total Tertiary sedimentary rock contained in these basins, about 20 percent is middle Miocene and older, and most of the rest is of Pliocene and Pleistocene age.[4]

The Tertiary basins of Indonesia represent three general types: extracontinental subduction basins, extracontinental median basins, and cratonic basins. Two kinds of subduction basins are petroleum productive: fore-arc and back-arc (foreland) basins. Fore-arc basins form between active arcs and the subduction boundaries of the major crustal plates. Sediments in these basins consist principally of outer shelf to neritic (deposited in 0–600 feet of water) shale, marl, and limestone often interbedded with volcanic rocks. Because of low heat flow, poor reservoir rock quality, and a lack of trapping structures, fore-arc basins are generally considered unfavorable habitats for petroleum.

Back-arc basins are located on the cratonic side of the volcanic arc

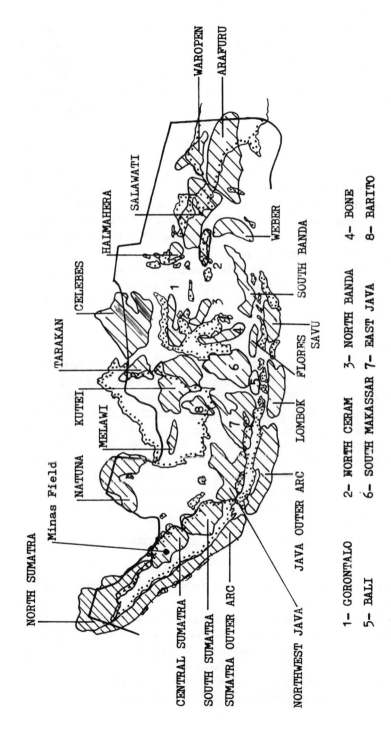

Figure 6.1. Map of sedimentary basins of Indonesia.

and are formed by the relative subsidence of cratonic plates behind active arcs that are superimposed on continental plate margins. They contain locally derived carbonate sediments or prograding wedges of sediments from the adjoining craton and abundant amounts of organic material. High geothermal gradients are common. Prolific petroleum discoveries have been made in Indonesian back-arc basins, and good potential remains for future discoveries.

Median basins are small basins of irregular profile that are formed in the "median zone" between a subduction zone and the basins of the craton. They consist of clastics, the more recent of which were deposited in massive delta complexes. Thus, delta basins have developed over the median basins. Interbedded shales provide both source and cap rock for sandstone reservoirs, and heat flow is sufficient to generate considerable oil. Prolific petroleum discoveries have been made in these basins and good potential remains.

Cratonic basins occur on the Sunda shelf and the Australian craton. These basins may contain excellent reservoir rocks, but the requisite structure and geothermal maturation of the organic matter is uncertain. Thus, although commercial discoveries have been made, considerable risk accompanies the significant petroleum potential of Indonesia's cratonic basins.

A third type of subduction basin is also found in the region. Inner-arc basins are formed behind active arcs that are superimposed on the oceanic plate margins. They occur in the same relative position as a back-arc basin, but without an adjoining craton; thus, they do not have a source for prograding sedimentary wedges. Inner-arc basins that contain thin and geochemically immature sedimentary sections may have only marginal petroleum potential. Some, however, do have relatively thick sections of sedimentary rock, and the deeper parts of the section may have been subjected to sufficiently high temperatures to generate oil. For the most part, these are located in fairly deep waters and thus have not yet been drilled.

The 28 Tertiary basins of Indonesia can be classified as follows:

- *Fore-arc basins:* North Ceram, Sumatra fore-arc, Java fore-arc, Lombok, and Savu
- *Back-arc basins:* North Sumatra, Central Sumatra, South Sumatra, Northwest Java, East Java, Salawati, and Waropen
- *Median basins:* Kutei, Tarakan, and Barito
- *Cratonic basins:* Natuna, Melawi, and Arafuru
- *Inner-arc basins:* Bone, Flores, Bali, South Banda, Weber, South Makassar, North Banda, Gorontalo, Celebes, and Halmahera

In Indonesia, six general types of geological units have been recognized as being favorable plays for oil exploration.[5] These are: transgressive clastics, regressive clastics, deltas, carbonate platforms, pinnacle reefs, and fractured igneous and volcanic rocks.

Transgressive clastic rock sequences contain coarse-grained, quartz-rich sandstones that were deposited by processes associated with a relative rise in sea level. These sand wedges, deposited in fluvial, near-shore, and deep-water environments (the last by turbidity currents), are usually good reservoir rocks. Where interbedded with shales that serve as source and cap rocks, folded into large anticlines, and thermally maturated, these multiple-pay reservoirs in transgressive clastic sequences have contained most of Indonesia's crude oil.

Regressive clastic rock sequences contain fine- to medium-grained quartz-rich sediments deposited by processes associated with a relative lowering of sea level. In Indonesia, regressive sequences form wedge-shaped bodies of sandstone that were deposited in near-shore to open-marine environments. Two or more potential sandstone reservoirs commonly are present and oil accumulations are in structural traps.

Deltas are vertically and laterally complicated, prograding sand-shale sequences that were deposited in response to the tectonic uplift of a river sediment supply area. The quartz sandstone bodies usually contain petroleum in medium- to thick-bedded pay zones in structural and stratigraphic traps. The cap rock and source rock are interbedded shales.

Carbonate platforms include a broad category of reefoid and bioclastic reservoirs, which have been developed upon a carbonate shelf. These reservoirs usually contain multiple pay zones with vugular or inter-granular porosity. Reservoir distribution is controlled by facies development and commonly modified by post-depositional tectonics. Seals are provided by shales or micritic limestones, and potential petroleum source rock occurs throughout the sequences. Pinnacle reefs occur seaward of the carbonate platforms as isolated vertical features in an intermediate zone between deep-basin shales and shallow-water carbonates. They are developed in response to a relative rise in sea level, commonly as a result of basin subsidence. Pinnacle reefs may also occur within the carbonate platform area in local deep basins. They have excellent reservoir characteristics, high production rates, and thick productive sections but are commonly restricted in area and contain only modest reserves. Tectonic activity has caused many pinnacle reefs to be uplifted, faulted, and tilted.

Petroleum is also produced from fractured igneous and volcanic rocks. The fractured reservoirs are sometimes capped by overlying fine-grained clastic sequences that also act as source rocks. Fractured basement reservoirs have been discovered by deep drilling on structural closures.

Significant Petroleum Basins

North Sumatra Basin

The North Sumatra back-arc basin is one of the medium-sized basins of Indonesia. The onshore portion of the basin is geologically well known from surveys and drilling. Petroleum production is from regressive clastic rocks and carbonate platforms of Miocene and Pliocene age. The basin's first oil was produced in 1885. The Rantau field, discovered in 1929, is the dominant oil field. Although well into decline, it still produces almost half of the basin's oil and contains over half of the basin's oil reserves. A number of other fields discovered before 1940 are also in decline or have been abandoned. These fields were shallow structural features that were discovered by surface geologic mapping. A major gas field, Arun, was discovered in 1971. The gas reservoir is deeper than the early oil fields and is in a middle Miocene reef limestone. This is the largest natural gas reserve in Asia, with proved reserves of 13.7 trillion cubic feet and additional inferred reserves of 1.4 trillion cubic feet. The gas is liquefied for export.

In the 1960s and 1970s exploration moved offshore and to the deeper portions of the basin where several discoveries have been made on the basis of gravity and seismic information. With many of the oil fields in decline, a continued decline in oil production from the North Sumatra basin is expected. Nevertheless, exploration continues, particularly offshore, with moderate prospects for discovery. Gas production is expected to increase with continued development of the Arun field. Prospects for gas discoveries are slightly better than for oil discoveries.

The North Sumatra basin produces only about 2 percent of Indonesia's oil and contains about 1 percent of the country's estimated oil reserves, but it contributes about half of Indonesia's marketed gas production and contains half of the nation's estimated gas reserves.

Central Sumatra Basin

The Central Sumatra basin is a small but very prolific back-arc basin that is well known geologically from outcrops and subsurface information. Its high well density and clustering of oil fields indicates an advanced state of exploration maturity. The sedimentary area of the basin appears to be restricted to shore. Relatively little geological work has been done offshore, which apparently offers poor petroleum prospects.

The oil fields are almost all anticlinal structures that were discovered by seismic surveys. Some of the larger shallow structures, including the giant and dominant Minas field, were discovered early in the exploration cycle on the basis of surface geologic mapping and shallow core drilling. The major producing geologic unit in the basin is the early Miocene Sihapas group, which contains fairly massive, fine- to coarse-grained sandstones interbedded with shales. The Sihapas is a transgressive deposit that includes reworked older sediments and contemporaneous clastics. The interbedded marine shales acted as both source and cap rock for the oil accumulations. The oil maturation process was aided by the generally high heat flows in the basin, which caused a considerable amount of oil to be generated from deposits of relatively small volume and moderate age at generally shallow depths.

The first oil discovery in the Central Sumatra basin was made in 1939, after which exploration was interrupted by World War II. The dominant world-class giant Minas field was actually found during the war by Japanese occupation forces when they began drilling a site that had been selected and prepared by a Western oil company. The Japanese used the drilling equipment at the site, which had been abandoned by the company at the time of the invasion of Sumatra. The Minas field is the largest known oil field in Southeast Asia, and the name Minas crude has become a universal term in world oil markets for waxy, low-sulfur crude oil.[6] The field can be considered Indonesia's greatest subsurface asset. Minas production began at an initial rate of 15,000 barrels per day and increased steadily until it reached 400,000 barrels per day in late 1969. It has since declined to about 335,000 barrels per day. Peripheral water injection was initiated in early 1970 to arrest excessive pressure declines. Injection averages about 300,000 barrels of water per day into about 28 wells on the southwestern and western flanks of the field.

Minas accounts for about 43 percent of current Central Sumatra basin oil production and 36 percent of the basin's estimated oil reserves. The Central Sumatra basin provides half of Indonesia's oil production; thus, Minas provides over 20 percent of the nation's oil.

The basin, though dominated by Minas, contains three other world-class giant fields (Bangko, Bekasap, and Duri) and two possible giants (Pematang and Pentani). However, production from all six of these major fields is in decline. (There are plans to steam flood Duri in an attempt to increase its production.) With the major fields in decline, Central Sumatra basin oil production is expected to decline in spite of a number of new, but smaller, fields being put into production. The prospects of additional discoveries are moderate, with any fields discovered expected to be smaller than the major fields now in decline. Even so, about half of Indonesia oil reserves are in Central Sumatra basin fields.

South Sumatra Basin

The South Sumatra basin is a moderate-size back-arc basin with a depositional history similar to that of the Central Sumatra basin. It is not, however, nearly as prolific in petroleum. Exploration began early in the basin, with at least two fields discovered before 1900. The earlier discoveries were in shallow, regressive, late Miocene to Pliocene sandstones. The larger fields, however, were found later in early Miocene transgressive clastics. There is also some production from carbonate platform rocks.

The dominant field in the basin is Talang Akar Pendopo, discovered in 1922. It contained about one-quarter of the basin's oil thus far discovered but is now well into decline and produces only about 2 percent of the basin's oil. Most of the fields in the South Sumatra basin are old and in a state of advanced decline, and many have been abandoned.

The South Sumatra basin accounts for about 2 percent of Indonesia's oil production and about 2 percent of the nation's estimated oil reserves. The basin has moderate prospects for additional discoveries of small fields, and oil production is expected to continue to decline.

Northwest Java Basin

The Northwest Java basin is a moderate-size back-arc basin that lies mostly offshore and includes the Sunda, Ardjuna, and Jatibarang sub-basins. Its depositional history includes a late Oligocene to mid-Miocene transgression. The transgressive sand reservoirs are found near the margin of the Sunda shield and become thinner toward the basin. Carbonate rocks were deposited on the Sunda platform in early Miocene time. A middle Miocene uplift of the Sunda shelf provided a source of clastic materials for the Northwest Java basin. This regressive phase of deposition was followed by a late middle-Miocene transgression during which carbonate rocks were formed. In late Miocene to Pliocene time, fine-grained regressive clastics were deposited over the entire Java Sea area.[7]

A variety of petroleum plays occur in the basin, including Oligocene-to Miocene-age transgressive and regressive clastic rocks, carbonate platforms, and pinnacle reefs. Perhaps the most unusual reservoir rock is in the Jatibarang field and is composed of Eocene to Oligocene fractured, volcanic, sandy tuffs. The numerous fractures are the controlling factor of the total porosity of these reservoirs.

Successful petroleum exploration in the Northwest Java basin dates only from the late 1960s, so the fields are relatively few and are not old by Indonesian standards. As they are mostly offshore, these fields had to await the development of the proper technology and favorable economic conditions before they could be exploited. The exception is the onshore Jatibarang field with its unusual reservoir of volcanic rocks. It contains 14 percent of basin reserves, making it the largest yet found in the basin. There are a number of large offshore fields under production and several new fields being developed, which indicates an increase in oil production. At the present time the basin produces about 14 percent of the nation's oil and contains about 15 percent of its total estimated oil reserves. A small amount of natural gas is also produced. The prospects for additional significant petroleum discoveries are good, especially offshore.

East Java Basin

Deposition in the East Java back-arc basin began in Eocene time with transgressive sandstones and platform carbonates. The Sunda shield became uplifted in late Oligocene, resulting in a regression of the sea and the deposition of regressive clastics. This was followed by a major transgression beginning in late Oligocene and ending in early Miocene. As transgression continued carbonates were deposited and, in the deeper central part of the basin, pinnacle reefs were formed. Following this transgression, a fall in sea level and a renewal of the erosion of the Sunda shield in the middle Miocene resulted in the deposition of regressive clastics across the region.[8] A final transgression occurred in late Miocene time, during which a carbonate complex was formed. This was followed by another regression of the seas and clastic deposition that continued into the Pliocene.

The East Java basin is of moderate size for an Indonesian back-arc basin, but it is not one of the better petroleum producers. The producing fields occur in two groups, one in the western portion of the basin and the other in the east, near Madura Island. Exploration in the basin began in the 1880s. Over half of the oil produced has come from the Kawengan field, discovered in 1894 and now about depleted. Many of the older onshore fields have been abandoned, and a productive offshore field has been shut in to conserve associated gas. Current oil production (less than 1 percent of total Indonesian production) comes mainly from a few old onshore fields that are far past their prime. East Java basin oil reserves are also less than 1 percent of the nation's total.

The recent offshore exploration in the basin has taken place in the eastern portion, near Madura Island. In this area the carbonate platform complex is the preferred target, but the transgressive and regressive clastics may also yield discoveries. These three plays have proved productive in the older fields of the basin. Without new discoveries, the East Java basin would appear to be in the final stages of decline.

The most prospective area is offshore from Madura Island, but even there prospects for commercial petroleum discovery are only moderate. The basin's current oil production decline is expected to continue.

Salawati Basin

The Salawati basin is a back-arc subduction basin located west of the Vogelkop portion of Irian Jaya. Tertiary deposition in this region was characterized by a late Cretaceous–Eocene transgression, an Oligocene regression, an early–middle Miocene transgression, and a late Miocene regression. During the middle Miocene the region was inundated and downwarped, which formed the Salawati basin.[9]

Beginning in the Miocene, basinal limestones interbedded with calcareous basinal shales were deposited. The shales are rich in organic material and may be the source rocks for much of the oil found in the basin. As the basin filled with sediments, shallower water depths allowed the formation of platform carbonates. The shallow-water environment provided favorable conditions for the growth of reef masses. When the region again subsided some of the reefs were able to keep pace with the deepening waters and grew vertically upward, eventually reaching heights in excess of 1,500 feet. These pinnacle reefs are middle to late Miocene in age. The off-reef areas were filled with a basinal shale, which provided excellent side and top seals to entrap the petroleum now found in the reefs.[10] The reef masses containing the oil fields in the Salawati basin exhibit a wide variation in size and shape. The Walio reef (a possible giant oil field) is about 12 miles long and 4 miles wide and dominates Salawati basin production. Sixty percent of Salawati oil comes from this field. It also contains 67 percent of estimated Salawati basin oil reserves. The basin accounts for about 5 percent of total Indonesian oil production and 5 percent of estimated reserves.

With the exception of Klamono, which was discovered in 1936 and is now in prolonged decline, the Salawati fields were found in the 1970s. These fields, however, have also recently declined, some by quite large amounts. Fortunately the declines experienced by Walio may be within operational variations. The significant declines and, in some cases, the abandonments of some Salawati fields may indicate that high production rates have caused rapid depletion of fields with rather limited reserves, as is sometimes the case with reef fields. Regional tilting and local uplift have caused the eastern reef belt platform area to be exposed at the surface, providing a large area of surface-water influx, which creates a strong water drive in most of the reefs. This drive, plus exceptional porosities, results in very high production rates for individual wells. If reserves are small, such production rates can rapidly deplete a field.

There have been a number of recent discoveries in the basin, some of which are offshore. The accumulations were all in reef reservoirs and, in some instances, were of gas and gas condensate. These discoveries may stabilize Salawati production, at least for a time. The prospects for additional discoveries are moderate for both the offshore and onshore portions of the basin.

Waropen Basin

The Waropen basin is a large back-arc basin located on the northern coast of Irian Jaya. It has a thick sequence of Tertiary sediments composed of fine clastics interbedded with platform limestones, and a transgressive wedge of coarse clastics occurs in the onshore portions. The Waropen basin contains large anticlinal features and other types of prospective petroleum traps. Little petroleum exploration has taken place, but the limited drilling that has been done encountered poor source rocks. Thus, in spite of the presence of oil and gas seeps, prospects for large commercial accumulations of petroleum are moderate at best.

Kutei Basin

The Kutei basin, a median basin in eastern Kalimantan, was formed in early Tertiary time and contains mostly clastic sediments. The maximum thickness of late Miocene and younger sediments is located offshore. These strata include a large delta system, extensive shallow marine shelf sediments, and deeper outer-shelf and bathyal deposits. The delta prograded eastward in the middle Miocene and reached its greatest extent during late Miocene and early Pliocene times. The middle Miocene tectonic event that uplifted central Kalimantan initiated the deposition of the massive Mahakam delta. This situation has, in general, continued to the present time.

In the Kutei basin oil has been produced from the deltaic complex, a vertically and laterally complicated sand and shale sequence deposited in response to the uplift of a river sediment supply area. This depositional environment results in multiple pay zones of medium- to thick-bedded quartz-rich sandstones that contain both oil and gas. The interbedded shale beds are both source and cap rocks. Oil has also been produced from regressive clastics, medium- to fine-grained quartz-rich deposits that were formed during a general shallowing of the sea. These clastics were deposited during the middle Miocene to early Pliocene regression also associated with the formation of the deltas.

The Kutei basin is large by Indonesian standards and contains two giant fields (Attaka and Handil) and a possible giant (Sanga Sanga). Exploration began onshore, along the coast, almost 100 years ago. Sanga Sanga was quickly discovered. It was a huge field and is now well into decline. Offshore seismic exploration began in the 1960s. Attaka and Handil were among the discoveries made.

The Kutei basin provides about 23 percent of Indonesia's oil production and contains about 23 percent of the nation's reserves. Handil produces about 46 percent of this total and Attaka adds an additional 28 percent. Thus, these two relatively recently discovered giants currently produce almost three-quarters of the basin's oil. Handil contains about 54 percent of the basin's estimated oil reserves. Associated gas is collected from some Kutei basin fields and is liquefied at the Bontang facility for export.

Delta basins often contain few giant fields. However, when one develops over a convergent margin, such as over a median basin, the underlying structure may influence oil accumulation so that more giants could be formed.[11]

A number of recent petroleum discoveries in the Kutei basin should help maintain current production levels. Commercial gas production should increase. The prospects for additional discoveries are good, especially offshore.

Tarakan Basin

The Tarakan median basin of Kalimantan is somewhat similar in its Tertiary sedimentary history to the Kutei basin. During an Eocene to early Oligocene marine transgression, the Tarakan basin received clastics from a landmass to the west. A short regression in the middle portion of Oligocene time was followed by a late Oligocene to middle Miocene transgression during which fine-grained clastics and massive shelf limestones were deposited. In the middle Miocene, Kalimantan was uplifted, initiating the deposition of thick, delta complexes in the Tarakan basin. One of these deltas is represented by the Meliat formation, formed by the deltaic sedimentation of the ancestral Sesiap River. The regression of the seas continued into Plio-Pleistocene time with the deposition of the regressive delta deposits of the Tarakan formation.[12]

In the Tarakan basin, oil has been produced from regressive clastics and deltaic deposits. Exploration began early in the basin, with fields discovered in 1900. The older fields are generally in decline and some have been abandoned. The Pamusian field, discovered in 1905, currently produces about one-third of Tarakan basin oil. The basin is a very minor producer, accounting for about 1 percent of Indonesian oil and about 1 percent of Indonesian oil reserves.

Recent exploration drilling has been deeper and several discoveries have been made. These deeper finds have been mostly gas. Tarakan oil production is expected to decline. New discoveries of oil will be necessary for the basin to match its former production levels. Overall prospects for additional discoveries are moderate, but in view of the relatively thick sedimentary section, offshore prospects may be somewhat favorable.

Barito Basin

The Barito median basin has a Tertiary history somewhat similar to the other two Kalimantan basins. In the Barito basin, oil has been produced from transgressive Eocene sandstones, regressive Miocene sandstones, Miocene reef limestones, and pre-Tertiary volcanics and extrusive igneous rocks. Exploratory drilling in the basin began in 1937. The Tandjung field, discovered during the first year of exploration, has completely dominated basin production. This field has declined by 88 percent since 1964 but currently produces all of Barito basin oil and contains all of the basin's estimated oil reserves. This amounts to less than 1 percent of total Indonesian oil production and oil reserves. Barito oil production will continue to decline, and the prospects for additional petroleum discoveries are poor.

Natuna Basin

The offshore Natuna cratonic basin is located on the Sunda shelf northwest of Kalimantan. The sediments consist mostly of clastics deposited in shallow marine or lacustrine environments. Organic debris is abundant due to the proximity of land areas, and thus potential source rocks are numerous. Geothermal gradients range from normal to high. Successful wells appear to be related to the higher geothermal gradients. The Gabus formation, believed to be of Oligocene age, is the oldest generally recognized Tertiary sedimentary unit. Thought to be deltaic in origin, it is the primary exploration objective in the Natuna basin.[13] The structure of the basin includes strike-slip and thrust faults and uplifts, including the Tenggol arch, all of which are thought to have influenced oil accumulation.

The Udang field, discovered in 1974, dominates Natuna basin production and reserves and was Indonesia's first frontier field. Commercial production began from the Udang-A platform in April 1979. This large offshore facility, erected in 290 feet of water, contains 14 wells. In 1980 Udang-B platform, with 15 wells, was set in place. Although other discoveries of oil and of a large gas field have been made recently, Udang currently accounts for all of the basin's petroleum production and estimated reserves. The Natuna basin contributes about 2 percent of total Indonesian oil production and contains about 2 percent of the nation's estimated reserves. Oil production is expected to increase with the further development of the Udang field and the exploitation of some of the newly discovered petroleum accumulations. Prospects for additional discoveries are good.

Arafura Basin

The Arafura is a cratonic basin located on and offshore from southern Irian Jaya. The basin is thought to contain carbonate rocks interbedded with shale, which may be good potential source rocks. Requisite structure and maturation of organic material may, however, be lacking. There has been little exploration in the basin, and prospects for petroleum discoveries are moderate.

North Ceram (Bula) Basin

In the North Ceram fore-arc basin, which stretches along the northern coast of the island of Ceram, sedimentation began in early Pliocene and continued until Pleistocene. The Plio-Pleistocene section unconformably overlies deformed beds thought to be of Triassic age. The oil has accumulated in shallow, late Tertiary sediments but may have been derived from the underlying Triassic beds that are locally shaley with a petroliferous character. The low geothermal gradients in the region also suggest that the oil may not have been generated in the shallow, late Tertiary sequence.

The North Ceram basin is very small and relatively shallow. It contains three very shallow oil fields that were discovered by drilling oil seeps. Two of the fields were extremely small and were abandoned by 1930. The third field, Bula, discovered in 1897, has accounted for almost all of North Ceram basin production. Currently Bula accounts for less than 1 percent of Indonesian oil production and reserves. In spite of what would appear to be limited prospects, there has been relatively recent exploration drilling in the basin. Two shallow wells were drilled in 1978; one was shut in and the other abandoned. North Ceram basin oil production is expected to continue its decline, and prospects for additional commercial petroleum discoveries are considered poor.

Bone Basin

The Bone basin is an inner-arc basin that lies mostly offshore between the south and southeastern arms of the island of Sulawesi. The basin is small and relatively young geologically, with most sediments being of late Miocene or younger age. It is also a region of low geothermal gradients. There have been three recent shallow gas discoveries onshore on the south arm of Sulawesi, along the flank of the Bone basin. The discoveries are in Miocene limestone reservoirs in anticlinal traps. The gas wells have been shut in for further development. The sediments in the central portion of the Bone basin reach 16,000 feet in thickness. Therefore, in spite of the low geothermal gradients and young sediments, it is possible that oil has been generated in the basin. However, much of the basin lies beneath more than 6,000 feet of water; thus, in the near future drilling can be attempted only around the basin's margins. Current prospects for commercial petroleum discoveries in the Bone basin are poor to moderate.

A summary of Indonesian petroleum basins is given in Table 6.1. The table illustrates the two general principles listed at the beginning of this chapter. First, most oil and gas is contained in a few large fields, but most fields are small. Of the 11 productive Indonesian basins, representing 5 basin types (back-arc, fore-arc, median, delta, and cratonic), 7 have had over half of their total cumulative oil production from a single field. In all of the basins, the largest field accounted for a very significant percentage of that basin's total cumulative production. The Central Sumatra basin has over 60 oil fields, many of which are quite small. However, in spite of 3 other giant fields and 2 potential giants, the Minas field currently produces 43 percent of the basin's oil and over 20 percent of total Indonesian oil. In basins with more limited production and fewer fields, 1 dominant field usually accounts for most of the production.

A corollary to the giant-field principle is that normally a single basin (or a few basins) will dominate a country's petroleum production. In the case of Indonesia, the relatively small Central Sumatra basin has experienced the proper combination of geological events to produce giant fields. Hence, this basin has accounted for over half of the country's total cumulative oil production, in spite of the fact that 10 other basins also produce oil.

Second, as exploration progresses, the average size of the petroleum fields being found decreases significantly, as does the amount of petroleum found per unit of exploratory drilling. The large fields are usually found first because they are large and therefore more obvious. Also, it is certainly good economics to drill the large structures first in the hope of finding a giant field. The table illustrates that the single field that dominates basin production in the Indonesian basins was often discovered relatively early in either of the first two phases of Indonesian exploration. The large fields discovered more recently are often offshore and thus were not available for drilling during the earlier years of exploration. This suggests that much of Indonesia's undiscovered oil will be either in smaller fields or in much more difficult environments (remote jungles or deeper offshore waters) and, therefore, will be more

TABLE 6.1   Summary of Indonesian Petroleum Basins

Basin	Percent of Indonesian Current Oil Production	Percent of Indonesian Current Oil Reserves

Central Sumatra basin
    Petroleum discovered—1939
    Largest field discovered—1944
        (Minas—53% of cumulative basin production)
	50	50

Kutei basin
    Petroleum discovered—1880
    Largest field discovered—1974 (offshore)
        (Handil—22% of cumulative basin production)
	23	23

Northwest Java basin
    Petroleum discovered—1969
    Largest field discovered—1969
        (Jatibarang—11% of cumulative basin production)
	14	15

Salawati basin
    Petroleum discovered—1936
    Largest field discovered—1973
        (Walio—38% of cumulative basin production)
	5	5

South Sumatra basin
    Petroleum discovered—1900
    Largest field discovered—1922
        (Talang Akar Pendopo—24% of cumulative
        basin production)
	2	2

Natuna basin
    Petroleum discovered—1974
    Largest field discovered—1974
        (Udang—100% of cumulative basin production)
	2	2

North Sumatra basin
    Petroleum discovered—1885
    Largest field discovered—1929
        (Rantau—57% of cumulative basin production)
	2	1

East Java basin
    Petroleum discovered—1888
    Largest field discovered- 1894
        (Kawengan—51% of cumulative basin
        production)
	1	1

Tarakan basin
    Petroleum discovered—1900
    Largest field discovered—1905
        (Pamusian—72% of cumulative basin
        production)
	1	1

Barito basin
    Petroleum discovered—1937
    Largest field discovered—1937
        (Tandjung—94% of cumulative basin
        production)
	less than 1	less than 1

North Ceram (Bula) basin
    Petroleum discovered—1897
    Largest field discovered—1897
        (Bula—99% of cumulative basin
        production)
	less than 1	less than 1

difficult and expensive to find and produce than the fields already exploited.

Oil Production

Indonesian crude oil production peaked in 1977 at 617 million barrels per year and then declined. By 1980 the decline had resulted in only 576 million barrels per year, although it was reversed in 1981 when production increased to 587 million barrels.[14] The factors contributing to the decline in production were the 1974–1978 exploration slump, which had a somewhat delayed impact on production, and the natural production decline in older large fields.

Indonesian oil reserves are currently estimated at 9.8 billion barrels.[15] The reserves/production ratio is, therefore, 17:1.[16] This can be compared with a reserves/production ratio of nearly 30:1 at the beginning of the 1970s and a ratio as high as 50:1 in the 1960s. Thus, Indonesia has not been replacing the oil reserves that have been produced.

Pertamina, the state oil company, forecast correctly that 1980 would be the bottom of the downward production spiral and that oil output would increase in 1981 as a result of the exploration boom that began in mid-1980. Pertamina also set a goal for 1984 of a record production of 668 million barrels of oil. Achievement of this production goal, given a constant reserves/production ratio of 17:1, will require a very successful exploration program. A mathematical analysis indicates that over 1 billion barrels of oil per year must be added to reserves in each of the next three years to reach the 1984 goal.

With an increasingly mature exploration situation in the major producing basins, it will be necessary to discover oil at a faster rate than in any sustained period during the 1970s in order to achieve the record production goal. Given the condition of maintaining a 17:1 reserves/production ratio, the oil production goal projected by Pertamina will be difficult to accomplish. Such a production goal would be possible to achieve temporarily by lowering the reserves/production ratio by infill drilling to produce existing reserves faster. Such action, however, will lead to even faster declines in the future as fields become depleted and competition for rigs from the development sector lessens exploration drilling.

In contrast to oil, the production of natural gas has increased since 1977 and is currently about 1.2 trillion cubic feet per year from a reserve estimated at 27.4 trillion cubic feet (a reserves/production ratio of about 23:1).[17] About half of the gas produced is utilized. Expansions are anticipated both in internal consumption of natural gas and in liquefied natural gas for foreign export.

Resource Assessment

An analysis of Indonesian oil-field production suggests that, at least in the near term, oil production will probably decline in 7 of the 11 producing basins. This includes the Central Sumatra basin, the country's most important producing basin. On the other hand, production is expected to increase in the Northwest Java basin and in the Natuna basin.

Most of the producing Tertiary basins have been rather thoroughly explored. Many of the large fields were discovered in the first phase of exploration, before 1930, and are now in decline. Prospects for new discoveries in explored basins, such as those in Sumatra, are moderate, but there is the expectation that smaller fields will continue to be found. It is unlikely, however, considering a century of exploration, that new giant fields will be discovered in these basins, and new giant fields are needed to help maintain oil production as the older giants decline.

Prospects for new discoveries of large fields are better for basins that have only recently been explored and where significant discoveries have been made. Such basins include the Northwest Java (especially offshore), the Kutei (also, especially offshore), and the Natuna offshore. The remote Waropen and Arafura basins, both little explored, may be viable prospects, although preliminary indications suggest that source rocks may be poor. Petroleum exploration is expected to continue in these two basins.

Given the advanced age of some of the largest fields in Indonesia and the maturity of oil exploration in many of the basins, it would appear that a major effort to maintain oil production in the older, large fields by secondary methods, and perhaps even tertiary methods where possible, will have to be coupled with the discovery of new fields of significant size to result in production increases that could be sustained over the next several years. This is not impossible, but it may not prove likely. All of the back-arc basins (except for the remote Waropen basin) are under development, and most have been rather thoroughly explored. Two cratonic basins, the Natuna and the Arafura, have undergone or will soon undergo exploration and drilling. The three median basins are also well into development. The basins that remain generally undrilled are either fore-arc or inner-arc basins, types considered to be somewhat less favorable to petroleum potential. Those inner-arc basins that have a thick sequence of Tertiary sediments and, therefore, possibly mature source rocks (e.g., the Bone basin, which has three shut-in gas discoveries; the Gorontalo basin; and the Halmahera basin) lie beneath relatively deep waters (3,250 to 16,500 feet), making exploration and production drilling very expensive.

TABLE 6.2   Conventionally Recoverable Undiscovered Petroleum Resources
of Indonesia

	Probability of Occurrence		
	95%	5%	Mean
Crude oil estimate (billion barrels)	5	35	16
Natural gas estimate (trillion cubic feet)	13	94	42

Source: Charles D. Masters and Joseph P. Riva, "Assessment of Conventionally
Recoverable Petroleum Resources of Indonesia," U.S. Geological Survey Open-
File Report 81-1142 (Washington, D.C.: Department of the Interior, 1981).

TABLE 6.3   Original (Ultimate) Recoverable Conventional Petroleum Resources
of Indonesia

Oil (billion barrels)		Gas (trillion cubic feet)	
Cumulative production	9.2	Cumulative production	3.6+*
Reserves	9.8	Reserves	27.4
Undiscovered resources (mean)	16.0	Undiscovered resources (mean)	42.0
Total oil	35.0	Total gas	73.0

*Likely understated; data not available.
Sources: See Table 6.2; also, *Oil and Gas Journal* (December 29, 1980, and December 28, 1981); *World Oil* (August 1981 and February 1982); *Petroleum
Economist* (August 1981); T. J. Stewart-Gordon, "Europe Sees Gas as Top Energy Source by 2000," *World Oil* (April 1982), pp. 181-187; and "USGS Issues
Reports on World Oil, Gas Potential," *Oil and Gas Journal* (December 7, 1980),
pp. 170-175.

Any assessment of undiscovered petroleum contains much uncertainty. Nonetheless, such assessments are of value in a general understanding of global resource availability. The World Energy Resources
Program of the U.S. Geological Survey recently assessed the conventionally recoverable petroleum resources of Indonesia.[18] The results of
the geologically based assessment are shown in Table 6.2. The low
value of the range is the quantity associated with a 95 percent probability
that there is at least this amount. The high value is the quantity
associated with a 5 percent probability that there is at least this amount.
The statistical mean is calculated to be the expected amount. Using
this information, it is possible to estimate the original (or ultimate)
recoverable conventional petroleum resources of Indonesia as 35 billion
barrels of oil and 73 trillion cubic feet of gas (see Table 6.3).

Thus, Indonesia is estimated to have discovered over half of its oil.
This suggests that much of Indonesia's remaining oil will be more

difficult to find and produce than was the oil already exploited. Remaining oil is likely to occur in smaller fields or in more difficult environments, such as remote jungles or deep water. To find and produce a portion of this remaining oil will require a much more intensive and expensive effort than has been expended to date, as more than half of the oil that has already been discovered existed in 13 giant and near-giant fields that were relatively easy to find and develop. Any new very large fields will most probably be found in little-explored areas rather than in areas that have been drilled for many decades. Smaller fields will continue to be found, but it will take many of these to offset production declines in the older, large fields.[19]

Production declines or even relatively level oil production for the next decade will have an adverse effect on Indonesia's ability to export oil. Domestic demand is currently relatively low (139 million barrels in 1980), but it is growing at a rate of more than 10 percent per year. This low domestic demand allows Indonesia, with its relatively modest oil resources, to be an important oil exporting country. If domestic oil demand continues to increase at its current rate, however, and other forms of energy are not substituted for oil, it is possible that in the 1990s Indonesia will need all of its produced oil domestically.

Natural gas has not been exploited as much as oil. A rather large resource is estimated to be available, which may offer Indonesia the opportunity to remain a petroleum exporter in the 1990s. Further expansion is planned for the existing liquefied natural gas facilities, and a petrochemical industry is being built to utilize gas.

Summary

Indonesia contains 28 Tertiary basins, mostly initiated by tectonic movements at the end of the Mesozoic era. The basins represent three general types: subduction, median, and cratonic. Three kinds of subduction basins occur: fore-arc, back-arc, and inner-arc. A combination of geological factors is necessary to create an accumulation of petroleum. In the Indonesian basins up to six kinds of geological environments have proved favorable for petroleum development. These are: transgressive sediments, regressive sediments, deltas, carbonate platforms, pinnacle reefs, and fractured igneous and volcanic rocks.

Indonesian crude oil production peaked in 1977 at 617 million barrels per year and then declined. By 1980 production had fallen to 576 million barrels, but the decline was reversed in 1981 with the production of 587 million barrels of oil. Indonesian oil reserves are currently estimated at 9.8 billion barrels. Thus, the present reserves/production ratio is 17:1. The reserves/production ratio at the beginning of the

1970s was nearly 30:1 and in the 1960s as high as 50:1. Therefore, Indonesia has not been replacing produced oil reserves with new discoveries. Pertamina, the state oil company, has set a goal for a record production of 668 million barrels of oil in 1984. Achievement of this goal will require a very successful exploration program. In an increasingly mature exploration situation in the major producing basins, it will be necessary to discover oil at a faster rate than in any sustained period of the 1970s to meet the production goal. It would be possible to increase production temporarily by lowering the reserves/production ratio by infill drilling to produce existing reserves faster. This would lead to even greater declines in the future, however, as fields become depleted and competition for rigs from the development sector reduces exploration drilling.

In contrast to oil, natural gas production has increased and is currently about 1.2 trillion cubic feet per year from a reserve estimated at 27.4 trillion cubic feet. Expansions are anticipated in both internal consumption of natural gas and in liquefied gas for export.

An analysis of Indonesian oil fields suggests that oil production will probably decline in the near term in 7 of the 11 producing basins. This includes the Central Sumatra basin, the largest producer. On the other hand, production is expected to increase in the Northwest Java and Natuna basins. Most of the producing basins have been rather thoroughly explored, and many of the large fields were discovered before 1930 and are now in decline. It is unlikely that very large new fields will be discovered in these basins, and such discoveries are needed to offset the declining older giant fields. Prospects for discoveries of new large fields are better for basins that have only recently been successfully explored.

A geologically based assessment indicates that there may be 16 billion barrels of undiscovered recoverable oil and 42 trillion cubic feet of undiscovered recoverable gas remaining in Indonesia. Thus, Indonesia may already have found over half of its oil, and the remaining oil will probably be more difficult to discover and develop than the oil already exploited. The remaining oil is likely to occur in the more difficult environments, such as remote jungle or deep water. To find and produce a portion of this remaining oil will require a much more intensive and expensive effort than was necessary in the past, as more than half of the oil already discovered existed in 13 giant and near-giant fields that were relatively easy to find and develop.

Production declines or even level production for the 1980s will have an adverse impact on Indonesia's oil exports. Domestic oil demand is still relatively low but is rising at a rate of more than 10 percent per year. This low domestic rate of consumption allows Indonesia to be an important oil-exporting country in spite of its relatively modest oil resources. If domestic oil demand continues at its present rate, and if other forms of energy are not substituted for oil, it is possible that in the 1990s Indonesia will need all of its oil production for domestic use.

Natural gas may offer Indonesia the opportunity to remain a petroleum exporter in the 1990s, as a rather large gas resource is estimated to be available. Further expansion is planned for liquefied natural gas facilities, and a petrochemical industry is being built to utilize gas.

Notes

1. Ooi Jin Bee, "Offshore Oil in Indonesia," *Ocean Management* (June 1980), pp. 51–73.

2. R. R. Steven, "Indonesian Activity Back to Full Force," *Offshore* (February 1980), p. 95.

3. Eugene B. Mihaly, "Is Adverse Oil Policy Shift Near In Indonesia?" *Oil and Gas Journal* (December 14, 1981), pp. 142–148.

4. G. S. Fletcher and R. A. Soeperjadi, "The Land of Plenty: Indonesia's 28 Tertiary Basins Hold 99% of Production," *Oil and Gas Journal* (August 22, 1977), pp. 150–156.

5. R. A. Soeparjadi, G.A.S. Nayoan, L. R. Beddoes, Jr., and W. V. James, "Exploration Play Concepts in Indonesia," *Proceedings of the Ninth World Petroleum Congress,* vol. 3 (London: Applied Science Publishers, 1975), pp. 51–64.

6. Kamal Madjedi Hasan and F. B. Langitan, "The Discovery and Development of Southeast Asia's Minas Oil Field," *Oil and Gas Journal* (May 22, 1978), p. 168.

7. Leslie R. Beddoes, Jr., "Hydrocarbon Plays in Tertiary Basins of Southeast Asia" (Paper delivered at the Offshore Southeast Asia Conference, SEAPEX Session, February 26–29, 1980), p. 3.

8. *Ibid.*

9. A. Pulunggono, *Recent Knowledge of Hydrocarbon Potentials in Sedimentary Basins of Indonesia,* Circum-Pacific Energy and Mineral Resources Memoir 25 (Tulsa, Okla.: American Association of Petroleum Geologists, 1976), p. 248.

10. Richard R. Vincelette and R. A. Soeparjadi, "Oil-bearing Reefs in Salawati Basin of Irian Jaya, Indonesia," *American Association of Petroleum Geologists Bulletin* (September 1976), pp. 1450–1452.

11. H. D. Klemme, "Petroleum Basins—Classifications and Characteristics," *Journal of Petroleum Geology* (October 1980), pp. 200–201.

12. Beddoes, "Hydrocarbon Plays."

13. E. P. Du Bois, Sr., *Synoptic Review of Some Hydrocarbon—and Potential Hydrocarbon—Bearing Basins of Southeast Asia* (United Nations Development

Program [UNDP], Technical Support for Regional Offshore Prospecting in East Asia [RAS/80/003], ROPEA—R.093), p. 23.

14. "Worldwide Oil and Gas At a Glance," *Oil and Gas Journal* (December 28, 1981).

15. *Ibid.*

16. The reserves/production ratio is a measure of the rate of petroleum production.

17. "Worldwide Oil and Gas at a Glance"; and "Natural Gas Reserves and Production Worldwide," *Petroleum Economist* (August 1981), p. 336.

18. Charles D. Masters and Joseph P. Riva, "Assessment of Conventionally Recoverable Petroleum Resources of Indonesia," United States Geological Survey Open-File Report 81-1142 (Washington, D.C.: Department of the Interior, 1981).

19. Joseph P. Riva, Jr., "Petroleum Prospects of Indonesia," *Oil and Gas Journal* (March 8, 1982), pp. 306–316.

7
The Petroleum Prospects
of the United States

Conventional domestic petroleum deposits occur in sedimentary basins throughout the United States. These basins are often dominated by a few massive geological structures that become traps for most of the basin's migrating petroleum. Because of their great size and anomalous geology, these deposits are usually found early in the exploration cycle. Of the more than 20,000 domestic oil fields that have been discovered, only about 275 are projected to ultimately produce more than 100 million barrels of oil. By the end of 1975 these national-class giants had produced about 60 percent of total domestic oil output and contained about 90 percent of proved domestic reserves. Eighty-one percent (223) of these fields had been discovered before 1950.

In 1980, about 3 billion barrels of oil were produced from about 527,000 wells in the United States. For production to remain at about this level for the next decade, 30 billion barrels of oil will have to be added to reserves or reserves will have to be drawn down below the current 9:1 reserves/production ratio. This amount is greater than current proved reserves or than inferred reserves. It will be very difficult to achieve such reserve additions in the mature provinces of the United States. Thus, the general decline in oil production that began in 1970 is expected to continue.

It also appears unlikely that sufficient gas will be discovered to prevent a general decline in domestic gas production during the 1980s. A short-term increase is possible through a further reduction in reserves/production ratios, which would result in a more pronounced decline toward the end of the decade. Although there is still hope of significant gas discoveries, especially in frontier areas, a continuation of the general decline in domestic gas production appears probable in the next decade.

Alaska

Alaska is located at the northern end of the American Cordillera, a continuous mountain system extending along the entire length of western North and South America. Therefore, Alaska is similar in geology to other regions of this long mountain system. It is one of the most volcanic and earthquake-prone regions of the world. In the northern part of the state, permanently frozen ground (permafrost) is common and constitutes an operational problem in petroleum development. Similarly, offshore, pack ice provides a development problem.

The two major producing basins in Alaska, at the present time, are the Arctic Slope basin (an extracontinental open downwarp basin) and the Cook Inlet basin (a back-arc subduction basin) (see Figure 7.1). The dominant field in Alaska is the Prudhoe Bay field on the North Slope. Discovered in 1968, this super-giant field was estimated to contain recoverable reserves of about 9.7 billion barrels of liquid petroleum and about 26 trillion cubic feet of gas, mostly in a gas cap above the oil. It consists of a succession of stratigraphic accumulations formed over a large structure as a result of the truncation of south- and southeast-dipping reservoir horizons by an Early Cretaceous unconformity. The main reservoir is the Triassic Sadlerochit sandstone.[1]

The trans-Alaska crude oil pipeline began operations in June 1977 and, with the help of a drag-reduction additive and additional pumping horsepower, is able to move crude oil at about 1.52 million barrels per day. This is near the optimum Prudhoe Bay production rate, which is usually given as around 1.5 million barrels per day. At the end of 1980 Prudhoe Bay had already produced 1.5 billion barrels of oil. Since construction has not yet begun on a gas pipeline, any produced gas is reinjected into the gas cap over the oil. Plans are being made to inject seawater beneath the oil in the reservoir to maintain pressure and thus increase ultimate recovery by an estimated 5 to 9 percent. Prudhoe Bay production is expected to begin to decline in the mid-1980s, and by the end of the decade may fall below 1 million barrels per day.

The other major petroleum-producing region of Alaska is Cook Inlet, a back-arc basin that contains about 60,000 feet of Mesozoic and Cenozoic sediments. The older Mesozoic rocks are mostly of marine origin, and the younger Tertiary rock sequence is largely nonmarine. Oil and gas seeps are known in the Mesozoic rocks, but petroleum production has come mostly from structural traps in the overlying Tertiary sequence. Oil was discovered in commercial quantities in the area in 1958 in the national-class giant Swanson River field on the Kenai Peninsula. Since that time, 6 oil fields have been developed (including the world-class giant McArthur River field) and about 18

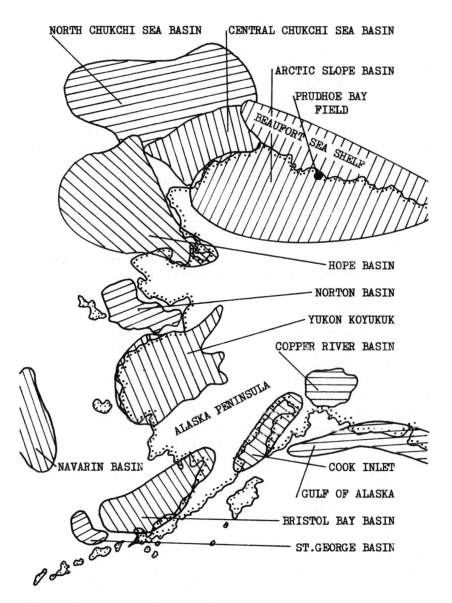

NORTH CHUKCHI SEA BASIN

CENTRAL CHUKCHI SEA BASIN

ARCTIC SLOPE BASIN

PRUDHOE BAY
FIELD

BEAUFORT SEA SHELF

HOPE BASIN

NORTON BASIN

YUKON KOYUKUK

COPPER RIVER BASIN

ALASKA PENINSULA

NAVARIN BASIN

COOK INLET

GULF OF ALASKA

BRISTOL BAY BASIN

ST.GEORGE BASIN

Figure 7.1. Map of sedimentary basins of Alaska.

gas fields have been located. Offshore drilling began in 1962 with the discovery of the North Cook Inlet gas field and the Middle Ground Shoal oil and gas field.[2] Offshore development followed rapidly. There are about 15 platforms that produce oil and gas from fields in the inlet.

Oil production from the region is about 100,000 barrels per day, but this is only half of the peak production achieved in 1970. For the past several years, Cook Inlet's production decline has exceeded 10 percent per year. Gas production averages about 580 million cubic feet per day. Part of the gas is exported as liquid natural gas to Japan and the remainder is utilized by local industry and for southern Alaska consumption. Several gas fields are shut in or are capable of greater production, but these will not be developed until markets are available.

The Upper Cook Inlet, where the producing fields are located, shows little additional promise for offshore petroleum discoveries, but there are a number of onshore areas yet to be explored on the Kenai peninsula and in the Chitina Valley–Copper River region.[3] Lower Cook Inlet so far has proved an expensive disappointment: the exploratory holes have been dry. Available data indicate that it has a thinner stratigraphic section and also less structural potential than the upper inlet.

Petroleum Geology

The Arctic Slope, also called the North Slope, includes an area in excess of 100,000 square miles, a significant portion of which is considered to have petroleum potential. Oil seeps along the Arctic coast have been reported since the early 1900s. In 1923, 37,000 square miles of the North Slope were set aside as Naval Petroleum Reserve No. 4, an area that has undergone considerable federal petroleum exploration. The average thickness of the sedimentary rocks in the North Slope open downwarp basin exceeds 16,000 feet. Older Mississippian to Jurassic sediments are overlain by organic-rich Lower Cretaceous shales that are thought to have been the source of the petroleum trapped in the super-giant Prudhoe Bay field. It is possible that some of the pre-Cretaceous rocks that contain petroleum at Prudhoe Bay may locally extend across the Barrow arch and under the Beaufort Sea. It is not expected that another Prudhoe Bay–size field will be discovered offshore, however, because the larger structures there appear to be complexly faulted, thus limiting the size of the prospective fields and making any accumulations hard to find. The exploration targets in the region will be not only the Permian-Triassic sands, but also Mississippian and Pennsylvania strata and the younger Cretaceous sands.[4]

Naval Petroleum Reserve No. 4 was transferred to the U.S. Department of the Interior in June 1977. Two small gas fields, which are "commercial" only because the gas is used locally, have been discovered in the reserve. Exploration in the reserve continues.

Other prospects on the North Slope include the world-class giant Kaparuk field, located about thirty miles west of Prudhoe Bay, with potential oil reserves estimated to be 1.3 billion barrels. Kaparuk production is scheduled to begin at about 80,000 barrels per day. By 1985, with the help of a waterflood, production is expected to have increased to about 250,000 barrels per day and to continue at that level for about a decade. The flow will raise total North Slope production, but this rise can be accommodated by the pipeline through the further utilization of drag-reducing additives.[5]

To the north of Kaparuk, a potentially commercial reservoir has been discovered at Milne Point and a second potentially commercial discovery has been made east of Milne Point at Gwydyr Bay. Exploration wells in both areas indicate substantial petroleum accumulations, but because of the remoteness of these Arctic locations additional confirmation wells are necessary to demonstrate commercial-size deposits. Other promising areas that may be developed include the Sag Delta, northeast of Prudhoe Bay, and Point Thomson, located about fifty miles east of Prudhoe Bay. Exploration drilling in these areas has encountered significant amounts of petroleum.

Lying between the Prudhoe Bay region and the discoveries in the Mackenzie River delta of Canada is the 18-million-acre National Arctic Wildlife Range. The available geological information indicates that several closed geologic structures exist in the range. Two are huge, rivaling Prudhoe Bay in size.[6] Leasing in the range is prohibited by law for at least five years, but seismic surveys are permitted. Discoveries of oil and gas on both sides of the range and the existence of the large geologic structures, which may indicate potential petroleum traps at depth, are considered favorable indications of the existence of potentially commercial petroleum accumulations within the range.

Alaska is a part of the North American lithospheric plate, which is overriding the Pacific plate in the vicinity of the Aleutian trench (see Figure 1.5). In the Aleutian Island arc area and the Bering Sea the basins are mostly of the subduction type. Median basins occur along the west coast and in the interior of the state. Pull-apart basins have been formed off the northwest coast.

The Hope basin lies beneath the Chukchi Sea, but a portion extends onshore and has been tested by drilling (see Figure 7.1). Since the wells were dry this basin has been somewhat downgraded in petroleum potential. Seismic surveys indicate that the basin is moderately deep with large basement faults. Its sediment fill is thought to be Late

Cretaceous and Tertiary clastics. Petroleum recovery from beneath the Arctic pack ice will depend on as-yet-undemonstrated technology.

North of the Hope basin, two deep Mesozoic basins underlie the Chukchi Sea. The North and Central Chukchi basins are divided by the Barrow arch, a subsurface structural ridge. The Central basin contains Mississippian-to-Jurassic sediments overlain by organic-rich Early Cretaceous shales considered similar to the source rocks of the Prudhoe Bay field. North of the Barrow arch, a thick sequence of bedded rocks, thought to range from Cretaceous to Tertiary in age, dip northward to the continental slope. Several large diapir folds occur in the section, some piercing the entire sequence up to the seafloor.[7] As in the Hope basin, the technology to produce petroleum under the pack-ice conditions of the Chukchi Sea remains to be demonstrated and will be necessary for exploitation of these basins.

The Norton basin is located between the Seward Peninsula and the Yukon delta. It consists of a series of Late Cretaceous down-dropped grabens separated by uplifted horsts.[8] The grabens contain the thickest sediment fill, up to 23,000 feet, while the sediment thickness over the horsts is generally thinner than 10,000 feet. The best reservoir potential is believed to be in late Miocene and younger rocks. Potential traps are structural closures produced by sandstones draped over the horsts.[9] Water depths in Norton Sound range from 25 to 150 feet. In the inner sound, where depths are less than 60 feet, artificial gravel islands would allow year-round drilling. In the deeper, open waters winter ice would limit drilling to the four- to six-month summer open-water season.

The Navarin basin is the largest of the Bering Sea basins (300 miles long by 160 miles wide). It is located between the Pribilof Islands and Cape Navarin in the Soviet Union. The northwest part of the basin lies within Soviet territory, although the exact boundary has not been resolved. The basin is a graben produced by tensional faulting along the continental margin and is filled with 35,000 feet of interbedded sandstone and shale of probable marine origin, ranging in age from Eocene to Recent. Seismic surveys indicate that the upper part of the Tertiary section (Miocene and Pliocene) contains potential sandstone reservoirs. Seismic reflection data also have revealed several hundred bright-spot anomalies that may be related to shallow gas deposits and deeper bright spots that could correspond to petroleum deposits. Possible traps include sediment drape over basement highs and fault closures.[10] Favorable seismic indications and recent Soviet discoveries onshore in eastern Siberia have resulted in evaluations of the Navarin basin as the most prospective of the Bering Sea basins. Operating conditions will be extremely difficult, however: drilling will be hundreds of miles from an operations base and must be done in the summer to avoid ice, but even summer storms are violent.

The St. George basin is located on the Bering Sea continental shelf between the Pribilof Islands and the Alaska peninsula. Lying under 300 to 500 feet of water, the basin is defined by seismic surveys as an elongated Tertiary sedimentary depression formed by a northwest-trending structural graben filled with more than 30,000 feet of essentially unfolded Eocene to Recent sediments. The rocks that appear to have the greatest petroleum potential are Miocene sandstone units, and the best source rocks appear to be the Eocene and Oligocene shales of the deep graben. Petroleum traps that may be present include basement structural highs on the basin margins and fault closures against the steep basin walls.[11]

The Bristol Bay basin is the most-explored and best-known basin on the Bering shelf. Seismic work has been under way in the region since 1966, and surface geology and exploratory drilling have been carried out along the southeast margin onshore. Some of the holes have contained minor oil shows. The basin is located on the north side of the Alaska peninsula and is one of the largest Tertiary basins around the Pacific rim. Marine seismic data indicate that the Bristol Bay basin contains a very thick section of gently folded, Tertiary marine and nonmarine clastic and volcanic rocks. Several structural highs are located at the margins of the basin and offer good potential for petroleum accumulations.[12] The Cenozoic rocks that appear to offer the best reservoir potential are the Miocene sandstones. The offshore portion of the basin lies under rather shallow water, 50 to about 400 feet, at the southern tip of the Bering Sea ice.

The Gulf of Alaska region contains a thick sequence of Tertiary-age continental and marine strata. Structurally, the basin is composed of east-west trending features, including faults, similar to those found onshore. One small abandoned onshore oil field, discovered in 1902, demonstrated that petroleum has been generated in the basin. Further onshore exploration was unsuccessful, but several structures were defined offshore by seismic surveys. The central gulf area was leased and 10 holes were drilled offshore at costs estimated at $15 to $20 million each. All were dry and had to be abandoned. The eastern portion of the Gulf of Alaska remains prospective. A large anticline about forty miles offshore is thought to contain strata that are older than those encountered in the nonproductive central gulf. This area has been leased and is scheduled to be tested.[13]

Onshore, the Yukon-Koyukuk area extends south from the Kobuk River to the Yukon River in the area of Koyukuk and Nulato and then continues southwestward to the coastal lowland. It comprises about

100,000 square miles. The region is underlain by predominantly Cretaceous rocks with exceedingly complex structure. This general structural complexity and the scarcity of favorable reservoir rocks and source beds appear to limit petroleum potential to areas underlain by shallow marine and nonmarine Cretaceous rocks.[14] A few exploratory wells have been drilled, apparently without success.

Production, Reserves, and Resources

The petroleum history of Alaska for the past decade is shown in Table 7.1.*[15] The very large increase in reserve additions in 1970 represents, primarily, the Prudhoe Bay reserves. Although discovered in 1968, the reserves were not proved (by additional drilling) and added to Alaskan reserve estimates until 1970. With oil reserves in Prudhoe Bay estimated at 9,400 million barrels, it is evident from the table that Prudhoe Bay dominates Alaskan oil. Over 88 percent of the oil discovered in the state in the past decade was in this single super-giant field. The significant production increases that began in 1977 are the result of the completion of the trans-Alaska crude-oil pipeline, which allowed production to begin from the Prudhoe Bay field. Prudhoe Bay accounts for over 90 percent of Alaska's oil production, which was 591 million barrels in 1980 from a reserve of 8,799 million barrels. Inferred oil reserves in Alaska are estimated to be 5,000 million barrels.[16]

The undiscovered recoverable oil resources in Alaska have recently been estimated by the U.S. Geological Survey.[17] These modified estimates are shown in Table 7.2. For the state as a whole, the Geological Survey has calculated the range of undiscovered recoverable conventional oil onshore as 2,500 (95 percent probability) to 14,600 (5 percent probability) million barrels and offshore as 4,600 (95 percent probability) to 24,200 (5 percent probability) million barrels. High and low probability estimates are not additive.

Table 7.2 indicates that the expected amount of oil yet to be discovered in Alaska is about 19.2 billion barrels (12.3 billion offshore and 6.9 billion onshore). This amounts to only two more Prudhoe Bays. The distribution of the undiscovered oil is of importance. A 70-million-barrel field (Umiat) has been discovered on the North Slope but is shut in because it is not yet commercially viable, indicating that an onshore field would have to have reserves of at least 100 million barrels (or more, depending upon location) to be developed. Offshore, this amount may be expected to be at least doubled for a field to be commercially viable, particularly in areas of Arctic pack ice. This would indicate that some basins may have few, if any, commercial fields. As

---

*Tables 7.1–7.117 appear at the end of the chapter.

giant fields are usually not numerous in any basin, development in Alaska's offshore areas will not be intense. Development activity cannot be expected to even approach that in the Gulf of Mexico, where several thousand offshore platforms have been constructed.

Prudhoe Bay will begin its decline in the mid-1980s and will be producing less than 1 million barrels per day in the early 1990s. Because of long lead times, replacement of this lost oil production will have to come from other North Slope prospects now being developed. It is possible, but not assured, that North Slope oil production can be held at current levels during the latter part of the 1980s, but production in the 1990s depends upon exploration success in the frontier areas, particularly offshore. This will require additional federal leasing and the commercialization of the technology necessary to develop oil fields under Arctic pack ice.

Ultimate Alaskan oil recovery is estimated at about 35.5 billion barrels (see Table 7.3). Alaska is a young petroleum province, with over half of its estimated total recoverable oil resource yet to be discovered.

In a recent report, the National Petroleum Council estimated that the offshore and Arctic Slope portions of Alaska may hold undiscovered recoverable oil resources ranging from 18 to 21 billion barrels.[18] These estimates are similar to those of the U.S. Geological Survey. The council stressed that basins listed as having low potential should continue to be studied because additional geological information in such frontier areas could still cause significant revisions (either upward or downward) in petroleum resource assessment. The final confirmation of a resource estimate can come only from exploratory drilling. The National Petroleum Council estimated (as did the Geological Survey) that the Beaufort Sea has by far the greatest petroleum potential of the Alaskan frontier areas.

Alaskan natural gas production history for the past decade is shown in Table 7.4. The large increase in gas reserves added in 1970 resulted from the addition of the 26 trillion cubic feet of gas in the Prudhoe Bay gas cap. As this gas is not currently being produced (because there is no gas pipeline), the reserves/production ratios are very high. Even without Prudhoe Bay gas reserves the Alaskan gas reserves/production for 1980 would be 33 : 1, illustrating the difficulty of getting Alaskan gas to market. Several Cook Inlet gas fields are shut in or are capable of greater production. Production in 1980 was only 0.213 trillion cubic feet from a reserve of 33 trillion cubic feet. Inferred natural gas reserves of Alaska (projected field growth) are estimated to be 5.8 trillion cubic feet.

Undiscovered recoverable conventional natural gas resources have been estimated recently for Alaska by the U.S. Geological Survey. Table 7.5 contains a modification (100.7 trillion cubic feet) of these estimates. The Geological Survey estimate of undiscovered gas resources for all of onshore Alaska ranges from 19.8 to 62.3 trillion cubic feet and for all of offshore Alaska from 33.3 to 109.6 trillion cubic feet (95 to 5 percent probability). These undiscovered gas resource estimates can be compared to those of the Potential Gas Committee, which recently estimated 44 trillion cubic feet (16 trillion cubic feet possible plus 28 trillion cubic feet speculative) for onshore Alaska and 93 trillion cubic feet (13 trillion cubic feet possible plus 80 trillion cubic feet speculative) for offshore Alaska.[19]

The National Petroleum Council recently estimated the undiscovered gas resources potential of the offshore and Arctic Slope areas of Alaska. The amount estimated was 109 trillion cubic feet of potentially recoverable natural gas and equivalent natural gas liquids. Of this total amount, 41 trillion cubic feet was estimated to be associated with oil and 68 trillion cubic feet was estimated to be nonassociated. Under the assumptions used in the analysis, 10 trillion cubic feet of the nonassociated gas was projected to be economically recoverable, but the study did not evaluate the more complex economics of associated gas.[20]

In general, Alaska is still a frontier area, but today there are better interpretations about petroleum limits and potentials. Recent seismic work and drilling have diminished several areas once considered very prospective, such as portions of the North Slope basin, the Gulf of Alaska, and the Lower Cook Inlet. On the other hand, there are several large areas such as the Chukchi Sea, the Norton basin, the Navarin basin, the Beaufort Sea, and the Arctic Wildlife Refuge that have not been drilled and could hold significant amounts of petroleum.

The ultimate Alaskan natural gas recovery is estimated as about 141.8 trillion cubic feet, as is shown in Table 7.6. Alaska has considerable gas potential, with more than twice as much estimated undiscovered gas as gas that has been discovered to date.

The Energy Information Administration no longer distinguishes between Prudhoe Bay natural gas liquids and crude oil, as they are commingled for pipelining and are not continuously metered during production. Therefore, the production of natural gas liquids was reported as zero for Alaska in 1980, compared to a reported production of 1 million barrels in 1979. Reserves of natural gas liquids in Alaska are around 400 million barrels, over three-quarters of which are in the Prudhoe Bay field.[21] Inferred reserves are proportional to inferred natural gas reserves, i.e., 70 million barrels of natural gas liquids. The U.S.

Geological Survey has estimated that the undiscovered recoverable gas liquid resources in Alaska are 3.4 billion barrels.

In summary, the petroleum remaining in Alaska includes about 33 billion barrels of oil, 3.9 billion barrels of gas liquids, and 139.5 trillion cubic feet of gas, as shown in Table 7.7.

Only 25 exploratory holes were drilled in Alaska in 1980. Fifteen were on the Arctic Slope, and 10 of these appeared to be dry (including 7 drilled under the federal program in the National Petroleum Reserve). There were 5 apparent petroleum discoveries on the Arctic Slope in 1980. Eight exploratory wells were drilled in the Cook Inlet basin and two in the Copper River region; all were dry. In 1980, 117 development and service wells were completed, primarily in the Prudhoe Bay and Kaparuk fields.[22]

Pacific Coast

The Pacific Coast region includes the states of Washington, Oregon, and California, both onshore and offshore.

Washington and Oregon

The basins along the coast and offshore in the Northwest are primarily subduction basins. Western Oregon and Washington and the adjacent continental margin occupy the site of a large, elongate Tertiary basin (the Western Oregon–Washington basin) (see Figure 7.2). The oldest rocks in most areas of this basin are early Eocene volcanics. Marine sandstones and siltstones of Eocene to Pliocene age are interbedded with volcanic and nonmarine sedimentary rocks. The eastern margin of the basin is covered by lava flows from the Cascade Range. The geologic history of the region is complex, having been complicated by changes in the environment of deposition and the emplacement of extensive igneous intrusions.[23]

To the west, the Columbia River basalt lava flows are thought to overlie sedimentary rocks that are potentially petroleum bearing. Although petroleum production in Washington is negligible, exploration occurred in 1980 and was centered in Jefferson, Clallam, and Kittitas counties. In Kittitas County a deep well is planned to penetrate 3,000 to 5,000 feet of basalt lava flows and test prebasalt sedimentary units that are believed to occur at depths of 15,000 to 20,000 feet. The basalt is folded into a nearly symmetrical anticline in the drilling area.[24]

The search for petroleum doubled in scope in Oregon in 1980, principally because of the discovery of commercial quantities of gas in the Columbia County Mist field in 1979. Seven wells have been completed in Eocene sandstones in this field, and production has increased to

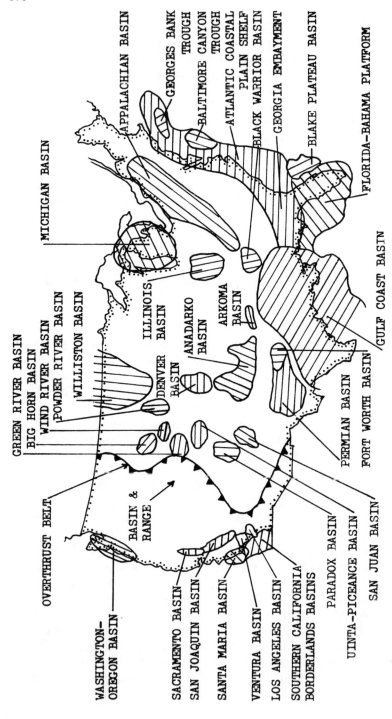

Figure 7.2. Map of sedimentary basins of the lower forty-eight states of the United States. In the Overthrust Belt, the barbs are on the overriding sheets.

over 20 million cubic feet of gas per day. However, the 18 exploratory and 3 extension wells drilled in Oregon in 1980 were all unsuccessful.

Drilling for petroleum in the Northwest has been quite unrewarding. An oil field found in 1957 at Ocean City, Washington, on the Pacific Coast produced for about four years and then was abandoned. There has been unsuccessful exploration drilling in portions of the Western Oregon–Washington basin since then, but the basin is still considered prospective. The U.S. Geological Survey recently estimated that the undiscovered recoverable oil in the basin ranges from 0 to 600 million barrels (statistical mean of 100 million barrels) on the continental shelf; from 0 to 1,300 million barrels (statistical mean of 200 million barrels) on the continental slope; and from 0 to 3,500 million barrels (statistical mean of 900 million barrels) onshore. Undiscovered recoverable natural gas is estimated to range from 0 to 2.8 trillion cubic feet (statistical mean of 0.6 trillion cubic feet) on the shelf; from 0 to 4.6 trillion cubic feet (statistical mean of 0.9 trillion cubic feet) on the slope; and from 1.9 to 19.2 trillion cubic feet (statistical mean of 7.7 trillion cubic feet) onshore.[25]

Several other areas of Oregon may have petroleum potential. There are several small sedimentary basins in the state, including the Vale and Harney basins, where exploratory wells have been drilled. Gas was discovered in the Vale basin a number of years ago but was used only locally.

California

California is an important oil-producing state, ranking behind only Texas, Alaska, and Louisiana in current oil production. Petroleum has been discovered in California in formations ranging in age from Jurassic to Pliocene, with occasional production from fractured basement schists. By far the largest amount of oil is produced from thick Tertiary sandstone reservoirs.

Petroleum Geology. The California fields are usually found in large anticlinal folds that parallel the mountain ranges. They are often cut by faults. Fifteen California fields are of world-class giant size. There are nonassociated gas fields, but much of California's produced gas is associated with oil fields. Many of the productive structural trends extend offshore. Thus, offshore exploration and development has been extensive, especially in the Santa Barbara Channel.

The onshore fields lie in an area between the Sierra Nevada, the coast ranges, and the sea. Two periods of orogeny (Late Jurassic and middle Pleistocene) combined to produce the mountains and their associated folds and faults. Large-scale movement at the boundaries of the region's lithospheric plates have produced some of the longest shear faults in the world (the San Andreas rift). The major producing basins of the state are the Los Angeles, Ventura, San Joaquin, and Sacramento basins (see Figure 7.2). These are non-arc extracontinental subduction basins that contain producing world-class giant fields. The Santa Maria basin is also a producing non-arc subduction basin but thus far is without a world-class giant field discovery.

Extensive oil seeps are common in the basins and result from the many folds and the absence of massive cap rocks. The preservation of large accumulations of oil in the California basins is due to the rapid deposition of overlying sedimentary cover, and their high productivity per unit volume of sediment results from the thickness and multiplicity of sandstone reservoirs and high geothermal gradients. The source rock is generally considered to be the highly organic Maricopa or Monterey shales, which are almost always in close contact with the reservoir sandstones.[26]

The first successful oil wells in California were drilled in the Ventura basin in 1875, an area of extensive seeps and asphalt deposits. The basin lies between the Santa Inez and Santa Monica mountains and extends westward beneath the ocean floor. There has been considerable successful petroleum exploration in the offshore portions of this basin. The petroleum accumulations occur in east-west–trending anticlines that are generally steeply folded and faulted. The Ventura-Rincon field, which extends offshore, is a world-class giant field with Pliocene sandstone reservoirs. It was discovered in 1916.

The early drilling in the San Joaquin basin was done on the basis of numerous surface seeps. The basin is about 250 miles long and 55 miles wide and contains a number of important anticlinal fields that resulted from the northwest-southeast folding. Other fields were formed by faults and unconformities. In general, the fields in the northern part of the basin contain gas and the southern fields include a number of major oil accumulations. The world-class giant fields in the San Joaquin basin, in order to size, are: Midway-Sunset, Kern River, Elk Hills, Kettleman North Dome (gas), Buena Vista, Coalinga, and Coalinga East Extension. These giant fields were discovered between 1890 and 1938. The producing horizons are Eocene, Miocene, and Pliocene sandstones.

The Los Angeles basin contains many very productive petroleum fields. It is bounded on the north by the Santa Monica Mountains, on the south by the San Joaquin Hills and the Santa Ana Mountains, on the east by the Repetto and Puente hills, and on the west by the sea. The basin contains a thick section of Miocene and Pliocene sediments that accumulated as the result of the rapid sinking of the Tertiary seafloor. The rocks were subsequently folded and faulted. Petroleum accumulations are densely scattered over the entire basin, perhaps the most prolific petroleum basin (on a sediment volume basis) in the world. Four principal producing trends occur, each associated with a major structural fold lineament that parallels the general northwest-southeast Tertiary fold belts of California.[27]

The Inglewood fault fields are composed of a line of fault and fold accumulations whose reservoirs are formed from thick Miocene and Pliocene sands. Included in this trend are the world-class giant Wilmington, Huntington Beach, and Long Beach fields. Wilmington and Huntington Beach lie partly offshore. Discovered in the 1920s and early 1930s, these three giant fields are faulted anticlines with Miocene and Pliocene sandstone reservoirs. Wilmington, the largest field in California, has produced about 2 billion barrels of oil.

The Whittier fault fields are a group of somewhat smaller oil fields that parallel the eastern mountain boundary of the basin. The oil accumulations usually occur on the down-thrown side of faults. The principal reservoirs are of Miocene age, but Pliocene sandstones are also productive. The largest of these fields is Montebello, a national-class giant discovered in 1917.

Parallel to the Whittier fault trend, but further to the west, is a minor structural lineament on which several large, faulted anticlinal fields occur. Santa Fe Springs, a world-class giant field, is the largest of these accumulations, which are collectively termed the Sante Fe pools. Discovered in 1919, Sante Fe Springs is an almost circular dome and has been one of the most prolific fields for its surface area in the world. It produces from a very thick section of Miocene and Pliocene sandstones.

The most westerly structural lineament, west of the Inglewood fault zone, is associated with the national-class giant Torrance oil field. Discovered in 1922, the field is in a gentle, southeastward-plunging faulted anticline. The reservoir rocks are of late Miocene and early Pliocene age.

The Sacramento Valley constitutes the northern half of the Great Valley of California and lies between the Sierra Nevada and the coast ranges. This is a gas province whose main reservoirs are of Late Cretaceous and Tertiary sandstones deposited in a non-arc subduction basin. Four tectonic episodes provided the structural and stratigraphic complexities that resulted in the gas traps.[28] By far the largest gas-producing field is the world-class giant Rio Vista, discovered in 1936. It is a broad, faulted dome with production from six Eocene- to Paleocene-age sandstones.

The Santa Maria basin, in which oil was discovered in 1902, contains the national-class giant Cat Canyon field along with several other smaller fields. Cat Canyon was discovered in 1908 by drilling an oil seep. The reservoir is a Miocene–Pliocene clastic in a faulted anticline. Trapping is aided by a facies change. The basin is located between the San Rafael and the Santa Inez mountains and extends offshore.

The Salinas basin, discovered in 1947, is also a non-arc subduction basin along the California coast. It contains the world-class giant San Ardo field, an anticline with a Pliocene-age sandstone reservoir.

The Southern California borderland is a complex of subduction basins, islands, banks, ridges, and submarine canyons with the edge of the continental shelf located about ninety-five miles from shore, beneath more than 13,000 feet of water. On the borderland, there are 7 major islands and 9 sedimentary basins. Many of the offshore basins are filled with great thicknesses of late Cenozoic marine sediments and may be similar to the Los Angeles and Ventura onshore basins. The near-shore basins are usually shallower and broader and contain mostly terrestrial sediments. An understanding of offshore petroleum potential relies on present knowledge of presumably analogous and better-known producing basins such as the Ventura, Santa Barbara, Los Angeles, and Santa Maria. Various studies of the rocks in the frontier offshore areas suggest that a number of shale or silty horizons, ranging in age from Late Cretaceous to late Miocene, contain sufficient organic matter to constitute potential source rocks for petroleum. Whether the deformational and thermal history of any of these formations has been appropriate for the genesis, migration, and trapping of petroleum is not known.[29]

The Santa Barbara Channel has the best offshore petroleum potential and contains over 30 producing oil and gas fields. Current production occurs from rocks ranging in age from Cretaceous through Tertiary. Major fields are the Don Cuadras and the Hondo, both of which are national-class giants estimated to contain over 200 million barrels of oil and/or equivalent gas.

The significant features of the northern and central California margin are a very narrow continental shelf, a very steep canyon-cut continental slope, a broad continental rise containing deep sea fans, and the occurrence of subduction basins. The continental margin in this area contains 6 structural basins that are extensions of onshore basins, with sediments of Tertiary age deposited upon Mesozoic igneous, meta-morphic, or sedimentary rocks.

Production, Reserves, and Resources. Onshore California petroleum exploration has reached a mature stage. There are only a few opportunities for new-field wildcats, and therefore exploration is directed mostly toward new pool tests and traps flanking known fields. In 1980, 403 onshore exploratory wells were drilled. In spite of the fact that 133 of these wells were completed as producers, only 5 new gas fields and 4 new oil fields were discovered.[30]

Offshore, 3 major fields have recently begun to produce. It is anticipated that these fields will add about 100,000 barrels per day to California production by the mid-1980s.[31] A recent federal lease sale in the offshore portion of the Santa Maria basin will allow early exploration of this new offshore area.

The petroleum history of California for the past decade is shown in Table 7.8. In spite of the advanced age of many of its giant fields, California's production has increased recently but has not yet equaled production at the beginning of the 1970s. The declining production in the Los Angeles basin and the coastal region has been more than offset by the increase in San Joaquin basin production. This increase is the result of significantly higher heavy oil production from such large heavy oil fields as Kern River, Midway-Sunset, and South Belridge. The increased production of heavy oil more than offsets oil production declines elsewhere in the state. The outlook for further increases in heavy oil production are good. Increased heavy oil production plus increased production from Elk Hills, the new offshore fields, and enhanced recovery projects in older fields such as Wilmington and Huntington Beach may, in the near term, continue to offset the general production decline of the older large fields and result in increasing California output. The recent increase in oil reserves is attributable to upward revisions in reserves in the big heavy oil fields, notably Kern River, Midway-Sunset, and South Belridge. In 1980, 360 million barrels were produced from reserves of 3,787 million barrels.

Inferred oil reserves in California are estimated to be 2,800 million barrels. Table 7.9 contains estimates of undiscovered recoverable oil resources in California by region. The total expected quantity is 7 billion barrels (3.4 billion offshore and 3.6 billion onshore). The table indicates that the San Joaquin basin, the Santa Barbara Channel, and the Santa Maria basin are estimated to have the best prospects for oil discovery. The best chances for the discovery of very large fields, however, are in undrilled areas offshore. The oil in the mature onshore basins will be found in the more numerous smaller fields, making it more expensive and difficult to locate and develop. The ultimate California oil recovery is estimated at about 32.4 billion barrels, as shown in Table 7.10.

California is a mature oil province in which over three-quarters of the recoverable oil estimated to occur has already been discovered. Nonetheless, a 7-billion-barrel undiscovered oil resource remains a considerable target for future exploration.

California gas production history for the past decade is given in Table 7.11. During the 1970s natural gas production in California declined by 55 percent. Reserves fell during the decade by over 1.6 trillion cubic feet, and 1.674 trillion cubic feet less gas was discovered than was produced. Gas reserves, however, increased to 5.010 trillion cubic feet in 1980. The increase was largely the result of upward revisions to the reserve estimates of existing fields, but new gas discoveries in the San Joaquin basin also contributed. The higher reserve estimates together with decreased production (only 0.299 trillion cubic feet) have resulted in California's highest gas reserves/production ratio of the decade (17 : 1).

Inferred gas reserves in California are estimated to be 3.4 trillion cubic feet. Table 7.12 contains estimates of undiscovered natural gas resources in California by region. The expected quantity (mean) is 10.9 trillion cubic feet.

Among the best prospects for new large gas fields are the undrilled offshore basins, such as the Santa Maria basin and portions of the Santa Barbara Channel. Exploration and development is in a mature stage in the Sacramento basin, but small gas fields are still being discovered. The San Joaquin and Ventura basins also have good potential, although the new fields will probably be relatively small in these mature basins. The ultimate recovery of California natural gas is estimated at about 48.1 trillion cubic feet and is shown in Table 7.13. California is a mature gas province, with an estimated 77 percent of its total content of natural gas already discovered.

In California, natural gas liquids production for 1980 was reported to be 8 million barrels with reserves of 120 million barrels.[32] Thus, natural gas liquids production averaged about 27 barrels per million cubic feet of natural gas production. On that basis, inferred natural gas liquids would amount to 92 million barrels. The undiscovered natural gas liquids resource has been estimated by the U.S. Geological Survey to be 500 million barrels. The estimated ultimate California gas liquids recovery is about 1.5 billion barrels, as shown in Table 7.14.

In summary, the estimated remaining petroleum in California includes about 13.6 billion barrels of oil, 0.7 billion barrels of gas liquids, and 19.3 trillion cubic feet of gas, as shown in Table 7.15. It can be seen from the table that California still contains a fairly substantial amount of petroleum. However, much of the total is in the form of proved and inferred reserves. The undiscovered petroleum will be in smaller fields or offshore and, in either case, will be much more difficult and expensive to find and produce than were the giant fields that were discovered early in the exploration history of the state.

Near-term prospects for increased oil production appear favorable because of increasing heavy oil production and production from new offshore fields. The prospects for increased natural gas production also may be considered somewhat favorable because of a higher reserves/ production ratio and new discoveries in the San Joaquin basin. Longer-term production, however, depends upon larger new discoveries. Estimates for undiscovered petroleum resources in the region are relatively modest and thus not likely to offset the decline in production in the older large fields in the longer term, in spite of recent enhanced oil recovery programs.

Rocky Mountains and Northern Great Plains

Thick sequences of Mesozoic sediments were deposited in the Cordilleran geosyncline, which covered large areas of the present states of Montana, Wyoming, Utah, Colorado, Arizona, and New Mexico and overlapped into North and South Dakota and Idaho. These sediments, and the underlying Paleozoic sequence, were subsequently uplifted and folded and faulted during an orogeny that began in the late Cretaceous epoch, and their erosion products were deposited as overlying Tertiary sediments in several intracontinental composite basins, the most important of which are: Paradox, Uinta, Piceance, San Juan, Wind River, Powder River, Big Horn, and Green River (see Figure 7.2). These are vertically composite sedimentary and topographic basins that, together with their peripheral belts of folded rocks, form the dominant structural features of the Rocky Mountain region.

The basins contain a large number of anticlines, many of which have steep dips and significant closure. Petroleum has been found at various stratigraphic levels in both Paleozoic and Mesozoic formations. Of these the Cretaceous sandstones have been the most prolific producers.

On the northern Great Plains, two interior basins are productive, the Williston and the Denver-Julesburg (see Figure 7.2).

Nevada and Western Utah

The Basin and Range–Great Basin region of Nevada and part of western Utah is considered to have significant petroleum potential, as several of the necessary geologic parameters for petroleum accumulation are present. However, it is possible that some petroleum deposits may have been destroyed by the severe and repeated structural deformation that has characterized much of this region.[33] Discoveries have proved elusive, but exploration continues. Only 5 fields have been discovered in the Great Basin, all producing from either Tertiary lake sediments or from fractured volcanic rocks. These oil accumulations occur in truncated fault traps or in sediments that drape over faulted structures. It took fifty years of exploration drilling and 77 dry holes before the

Eagle Springs field was discovered in 1954 and another twenty-two years and 100 more exploratory holes to find the next small field.

Exploratory work began in the Great Salt Lake in the late 1960s, prompted by numerous seeps of heavy oil. Several offshore wells were drilled and two recovered significant amounts of heavy, sulfurous oil. The reservoir is a fractured basalt, but about 15,000 feet of Tertiary sediments are believed to occur beneath the lake. They may contain both potential source and reservoir rocks, thus making the sequence petroleum prospective. Preliminary drilling has indicated potential reservoirs in both basalt and sandstone.[34]

Many of the exploratory wells drilled in the Great Basin–Basin and Range region have recorded good indications of the existence of petroleum. These small amounts of oil and gas have occurred in Tertiary and Paleozoic formations. Potential source rocks are thought to be Cretaceous and Tertiary lake deposits. Some Paleozoic structural prospects are also thought to occur. In addition, Paleozoic reefs, which may contain petroleum, are believed to be present.[35]

The Overthrust Belt

The Cordilleran Overthrust Belt of the western United States and Canada extends southward from Canada into Montana, Idaho, Wyoming, and Utah; westward to Nevada and southeastern California; and then eastward into Arizona and New Mexico (see Figure 7.2). The overthrust system is a complex of mountain ranges and basins evolved from a thick, mainly marine, series of sediments deposited in the Cordilleran geosyncline that persisted from late Precambrian to the middle of the Mesozoic on the western margin of the ancestral North American continent. Beginning in about the middle of the Mesozoic era, a series of orogenic episodes uplifted, compressed, and extensively deformed these sediments. The tectonic activity resulted in massive sheets of rock folded, faulted, and thrust eastward in an overlapping fashion. The total horizontal shortening across the thrust belt amounts to about 65 miles, or about 50 percent of the original sedimentary section.[36]

Major petroleum production from the Overthrust Belt occurs where overthrust sheets have moved eastward onto thick, marine, organic shales of Cretaceous age, as in southwestern Wyoming, northern Utah, and Alberta, Canada. The marine Cretaceous shales are thought to be the source rocks from which the petroleum was derived. The oil and gas were expelled from the Cretaceous source rocks and moved upward to be trapped in reservoir rocks in the upper plates of the complex thrusts. The reservoir rocks in these overlying thrust plates are mostly older (Triassic and Jurassic sandstones and limestones) than the Cretaceous source rocks but occur above them because of the overthrust faulting.

The Overthrust Belt is one of the most difficult and costly onshore regions in which to explore for petroleum. Before the deeper exploration could take place in the region, oil and gas prices had to rise, and improved seismic technology, including computer data processing, had to become available. Overthrust drilling is generally deep; some wells have been drilled below 18,000 feet. Down-hole environments at such depths are often highly toxic and corrosive. The presence of salt and the complex geometry of the rock layers may impede drilling and deviate holes, causing wells to cost as much as $8 million each.[37]

Due to these factors, the field generally credited with initiating deep overthrust exploration (the national-class giant Pineview in northeast Utah) was not discovered until early 1975. Pineview production comes from Jurassic and Cretaceous sandstones and limestones. Prior to the discovery of Pineview, about 500 relatively shallow dry holes had been drilled in the area. Since the discovery, a concentrated exploration effort has been underway in the Utah-Wyoming portion of the Overthrust Belt, with 175 exploratory holes resulting in the discovery of 16 fields. Current production from the overthrust belt is about 8.8 million barrels of oil and 0.024 trillion cubic feet of gas per year. However, gas production is limited by a lack of pipeline connections for sweet (low-sulfur) gas and the lack of processing plants for sour (high-sulfur) gas. Petroleum reserves have not yet been completely defined, as several fields are still not in production, but the region is estimated to contain proved and inferred reserves of 4 to 6 trillion cubic feet of gas and about 500 million barrels of oil.

A potential world-class giant field is the East Anschutz Ranch gas and condensate accumulation on the Utah-Wyoming border. It was discovered in 1979 and produces from a Jurassic sandstone reservoir.[38] Two other Wyoming overthrust fields of national-class giant size are Painter Reservoir and Whitney Canyon–Carter Creek.

The overthrust and disturbed belt of Montana is a geologically complex folded and thrust-faulted region with reported petroleum occurrences and potential source and reservoir rocks. The principal thrusts have large displacements that may conceal major structures. Due to the structural complexity, a thorough understanding of the geometry of the structures to be tested is necessary in order to properly locate a well intended to test deep potential reservoirs. Two fields have recently been discovered in the Montana disturbed belt. Oil and gas production in these fields is derived from Mississippian- and Cretaceous-age reservoirs. Although the size of these two fields has not as yet been determined, their discovery encourages continued exploration in the region.

South of the Utah-Wyoming portion, the Overthrust Belt is characterized by flat-lying thrusts and closely compressed folds. Volcanism is much more intense and, along with younger block faulting, may have contributed to the destruction of some southern thrust-belt traps. Source rocks and reservoir rocks are most likely of Paleozoic age, but potential traps would be much younger (late Mesozoic). A greater structural complexity will make geophysical techniques more difficult to apply than in Wyoming and Utah. Although indications of petroleum have been observed in test wells, commercial production has not yet been established in the Overthrust Belt south of the Utah-Wyoming discoveries.

Montana

A large number of oil and gas accumulations have been found in Montana. These occur in widely separated locations and are related to independent tectonic features. The most important structural trend in Montana is the Sweetgrass arch, a regional uplift that originates in the Little Belt Mountains and plunges northwest across the Canadian border into Alberta. The major petroleum fields of Montana are associated with this uplift, many of them accumulations in lenticular sandstone traps of Early Cretaceous age. Cut Bank, discovered in 1929, contains oil in Cretaceous stratigraphic traps on a plunging anticline. It is a national-class giant and the largest oil field in Montana. The Sweetgrass arch area is currently undergoing extensive drilling, with success ratios of about 25 percent. The favorite exploration targets are shallow Cretaceous gas sands; Devonian rocks also have been successfully tested.[39]

In the north-central part of the state several small petroleum fields occur that are associated with folds adjacent to the Sweetgrass Hills and Bearpaw Mountains. The central fields in Montana occur in association with the Big Snowy uplift. The principal reservoirs of these fields are Cretaceous sands interbedded with shales. Several Jurassic and Pennsylvania age stratigraphic accumulations have also been found in central Montana.

In the southeastern part of the state, in the Powder River basin, the national-class giant Bell Creek field is located. It is the second-largest field to be discovered in Montana. Other fields have been found in the Montana portion of the Williston basin, where oil has been discovered in Paleozoic and Mesozoic sediments. The Montana part of the Williston basin is currently experiencing a high level of exploratory drilling. The success ratio for 1980 was 23 percent, and 41 of 46 Williston discoveries were made in the Montana portion.[40] The primary exploration target was Ordovician in age, but Mississippian and Devonian accumulations were also found.

North and South Dakota

The Williston basin has been the locus of the North Dakota oil industry since 1951. It is a large interior basin that contains about 8,000 feet of Paleozoic sediments (primarily carbonates), 5,500 feet of Mesozoic sediments (mostly marine clastics), and 2,500 feet of Tertiary sediments (continental clastics). During most of early Paleozoic time, the Williston basin subsided, but various shorter episodes of uplift and erosion also occurred.

During the first phase of exploration (in the late 1950s) attention was concentrated in the areas around the large structures associated with the Nesson and Cedar Creek anticlines. Accumulations of oil were discovered in Mississippian reservoirs along both structures, including the national-class giant Beaver Lodge field, which was found by seismic surveys on the Nesson anticline.

Exploration technology improved considerably during the 1960s. Multifold shooting and advances in digital recording and processing permitted seismic evaluation of the deeper and more remote portions of the basin, and advances in well logging (particularly formation density logging) made possible a better understanding of the complex carbonate reservoirs.[41]

With such technological advances, it became possible to test Ordovician structures in the more remote parts of the basin. Another refinement, which occurred in the 1970s, was the development of a formation fracture identification log for interpreting fractured reservoirs. The discovery of the Red Wing Creek field in 1972 initiated the current phase of exploration and development. This discovery, with its relatively highly productive wells and unusually thick reservoirs, caused increased interest in Williston basin exploration. As a result of this renewed activity, two national-class giant fields (the Mississippian Mondak and Little Knife) were discovered in 1976. The fields are stratigraphic with some structural control, and unlike previous Mississippian stratigraphic traps their petroleum occurs in dolomite rather than in limestone reservoirs.

There also has been substantial production from recent discoveries along the Billings Nose trend from the lower Paleozoic sequence. The Billings Nose is located in the deep Williston basin southwest of the older Nesson anticline fields. Among the discoveries being developed in this area is the Big Stick–TR–Tree Top group of fields, which may eventually be found to be parts of a single large field. Production from this group is from carbonate reservoirs ranging in age from Ordovician to Mississippian.

The results of increased drilling activity in the Williston basin have been very successful. Reserves of individual fields are not outstandingly high by world standards, but the aggregate reserves of all the new discoveries are substantial. Some of the wells have been tested at rates in excess of 2,000 barrels of oil per day, but many reservoirs are of small lateral extent. Gas production also has been found in the deeper parts of the basin (below 13,000 feet). Exploration activity has advanced to the outer margins of the basin where large areas remain unevaluated.

The small amount of petroleum produced in South Dakota comes mostly from the outer margins of the Williston Basin. In 1980, however, the two South Dakota exploratory wells that discovered new fields were drilled on the margins of the Powder River basin. They both produced from shallow formations (Pennsylvanian and Jurassic).

Wyoming

Wyoming is the major oil-producing state in the Rocky Mountain–northern Great Plains region. It contains four major basins, all of which are of the intracontinental composite type. There are a number of large fields in these basins. The fields usually occur in anticlinal structures whose main reservoirs are Cretaceous sands. Stratigraphic traps, ranging in age from Paleozoic to Tertiary, have also been found, some of which are accumulations along the flanks of the anticlinal folds.

The Big Horn is the most oil productive of the Wyoming basins. Its largest field is Elk Basin, an elliptical anticline accumulation on the Montana border. Production is from Cretaceous sandstones and Mississippian limestones. Discovered in 1915, Elk Basin is a world-class giant field, but due to its advanced stage of development is currently not the leading Big Horn basin oil producer. The national-class giant Oregon basin field, discovered in 1912, produces almost twice as much oil. It is located on the west side of the Big Horn basin in an anticlinal fold and produces from Permian limestone, Pennsylvanian sandstone, and Mississippian limestone. Other Big Horn basin national-class giants include Grass Creek and Byron. Recent exploration drilling has concentrated along the flanks of the basin, but there has been some increase in deep drilling in the central portion.

The Wind River basin, in central Wyoming, contains a large number of oil fields that are located mostly in peripheral anticlines. There is, however, a deep gas potential in the more central parts of the basin. Production is from many different Paleozoic and Mesozoic reservoirs. Much of the most recent exploratory drilling activity in the Wind River basin has been in the vicinity of the deep Maden gas field, a major field with a depth of about 20,000 feet.

The Green River basin of southwestern Wyoming is the largest of the Wyoming basins and contains numerous oil and gas fields. The

Lost Soldier field, on the eastern side of the basin, is a national-class giant oil field that was discovered in 1916. It is an elliptical dome with 3,000 feet of closure. Production comes from several reservoirs of Pennsylvanian, Jurassic, and Cretaceous age. There are a number of other smaller eastern fields that are also in anticline-type folds. Similarly, along the western side of the basin, several petroleum accumulations exist in structural and stratigraphic traps. Recent exploration has concentrated on gas production from the Green River basin. The principal reservoirs for gas are Cretaceous sandstones in the deep eastern portions of the basin. Abnormally high pressure gradients occur in these formations, caused by the generation of natural gas at depth from carbonaceous shales and coals. The overpressured condition of the reservoirs allows more gas to be packed into the available reservoir pore spaces. Recent drilling has resulted in the discovery of a number of deep reservoirs of overpressured gas in the basin. Reserves in these reservoirs have been estimated at about 2.2 trillion cubic feet.[42]

The Powder River basin is located in the northeastern corner of Wyoming and the southern corner of Montana and contains many important oil fields. It has a large volume of effective source rocks of Cretaceous age that expelled many billions of barrels of oil during Tertiary time. This oil has undergone various kinds of alteration as it migrated and accumulated according to the current structural configuration of the basin. Permian oil also occurs, having migrated into the basin during Jurassic time from its main source to the west. This migration and expulsion of oil has influenced the formation of several large oil fields.[43]

Salt Creek, discovered in 1906, remains the largest field yet found in Wyoming. It is a large anticline on the western margin of the Powder River basin. A world-class giant field, it is divided into three main sections by a complicated system of faults. The central Salt Creek dome is the main producer, with a closure of 1,600 feet. The southern extension of the field is the Teapot dome, a fault-block structure limited on the north by reverse faults. Salt Creek, a very old field, is estimated to already have produced about 95 percent of its recoverable reserves. Production is derived from Mississippian, Pennsylvanian, Jurassic, and Cretaceous reservoirs. Lance Creek, discovered in 1918, is a national-class giant field located in a large asymmetric anticline with reservoirs of Cretaceous age. This field is also nearly depleted. Another national-class giant oil field in the basin is Hilight, discovered in 1969. It is a stratigraphic trap located in the northern part of the basin.

Over half of the exploratory wells drilled in Wyoming in 1980 were in the Powder River basin. New field wildcat discoveries were primarily of oil in Cretaceous and Permian sandstone, but there also have been discoveries of deep gas in the basin.

Utah

Two areas of oil activity in Utah that have not been previously discussed are the Uinta basin in the northeast and the Paradox basin in the southeast. The Paradox is a large intracontinental composite basin. The Aneth field, a national-class giant discovered in 1956, is the dominant field of the basin. The oil in the Paradox basin is mainly found in algal, coral, and oolitic limestones of Pennsylvania age, but other Paleozoic formations may also be prospective. The fields are primarily in stratigraphic traps. The recent discovery of the Bug field, also in Pennsylvania rocks, has revived interest in the exploration of the Paradox basin.

The Uinta basin is also an intracontinental composite basin. The dominant field is Greater Red Wash, a national-class giant that is a combination structural-stratigraphic accumulation of both oil and gas in Cretaceous sandstones. Drilling activity in the Uinta basin is currently at a relatively low level, but there has been a recent discovery of a gas field.[44]

Colorado and Nebraska

The two main basins in Colorado are the Piceance in the west and the Denver-Julesburg in the northeast. The Piceance is an intracontinental composite basin, and the Denver-Julesburg is an interior basin. The oil fields of Colorado are comparatively small, with the exception of the Rangely field, a world-class giant located in the Piceance basin. Discovered in 1902 as a surface-expressed anticline, it produced shallow Mesozoic oil for many years before the discovery of deeper Paleozoic reservoirs in 1933. At the present time, however, most of its original reserves have already been recovered. Current drilling in the Piceance basin has been concentrated on gas in Cretaceous sandstones.

Most of the current exploratory drilling in the state is in the Denver-Julesburg basin, where the main exploration targets are shallow Cretaceous gas and petroleum accumulations in channels and sand bars formed during Cretaceous time. The basin extends into Nebraska, where production is mainly from Early Cretaceous horizons. Drilling activity increased in 1980, particularly in the deeper parts of the basin. The number of exploratory successes in Colorado has increased, due partially to a greater number of marginal well completions in response to increasing petroleum prices.[45]

Western New Mexico

The San Juan basin of northwestern New Mexico is an intracontinental composite basin containing a number of significant gas accumulations. The reservoirs of these mainly stratigraphic gas fields are Cretaceous sandstones. Relatively small oil accumulations occur in shallow Cretaceous sandstones, and deeper Pennsylvanian petroleum deposits also have been found. Current exploration includes the testing of Paleozoic horizons, and a considerable amount of development drilling is under way in existing gas fields.

Production, Reserves, and Resources

The petroleum situation in the Rocky Mountain–northern Great Plains region can best be put into perspective with a historical sequence of reserves and production on a state-by-state basis, as shown in Tables 7.16 through 7.21.

Although Wyoming is the leading oil-producing state in the region, production has declined by 19 percent over the last decade. Reserves have also declined, as 104 million more barrels of oil were produced than were added to reserves. Reserve additions improved in 1980, and as the large fields in the Wyoming portion of the Overthrust Belt are further developed additional short-term increases in reserves and production can be expected. The largest Wyoming fields were discovered before 1930. Oil production in 1980 was 125 million barrels from a reserve of 893 million barrels (see Table 7.16).

Colorado oil production has increased during the past decade at the expense of reserves, which were drawn down by 56 percent, as 226 million more barrels were produced than were added to reserves. Thus, reserves/production ratios have decreased significantly and production is at a near-maximim rate from very mature fields. Reserve additions recently increased, but the largest additions were added in 1933 with the discovery of deeper Paleozoic reservoirs in the Rangely field. Oil production in 1980 was 32 million barrels, and reserves were estimated at 175 million barrels (see Table 7.17).

North Dakota production has increased significantly with the relatively recent discovery of several large fields. Reserves, however, have declined by 25 percent and reserves/production ratios are very low. This may be due to an incomplete early assessment of the size of the new fields, which would indicate a growth in reserves as these new fields are developed. The low reserves/production ratio (5 : 1) indicates that the older fields are being produced at a near-maximum rate and that any further increases in production will have to be derived from expected reserve growth or from the discovery of new fields rather than from more rapid exploitation of oil fields. Oil production in 1980 was 38 million barrels from a reserve of 177 million barrels (see Table 7.18).

During the past decade Montana's oil production declined by 34 percent and reserves declined by 42 percent. In 1980, however, a number of Williston basin discoveries were made, making it a good year for reserve additions. Thus, production may stabilize or even increase slightly in the near term. Oil production in 1980 was 29 million barrels from a reserve of 160 million barrels (see Table 7.19).

Utah has experienced falling oil production since the mid-1970s, but reserve additions can be expected to grow in the near future due to overthrust discoveries. Near-term production should also increase as these new large fields are developed. Oil production in 1980 was 25 million barrels from a reserve of 142 million barrels (see Table 7.20).

Western New Mexico produces little oil. Its fields are mostly in a mature stage of development, and thus reserves/production ratios are low. Reserve additions were up in 1980. Oil production in 1980 was 4 million barrels from a reserve of 21 million barrels (see Table 7.21).

The recent history of oil production in the Rocky Mountain–Great Plains region as a whole is shown in Table 7.22. The region experienced generally declining reserves over the past decade, as 565 million more barrels of oil were produced than were added to reserves. Reserves increased in 1980 to 1,568 million barrels in response to increased drilling. Reserves/production ratios declined through the decade, and production also generally declined but increased in 1980 to 253 million barrels. The increases in reserves and production in 1980 are the result of significant discoveries in the Overthrust Belt and the Williston basin. In both of these areas, improved exploration technology and higher oil prices have recently allowed the testing of deeper and more subtle traps. The opening of such "frontier" prospects, which could not be drilled before, has increased production in a region in which the older fields are in decline. It can be expected that, in the near term, this moderate increase will continue as the large frontier discoveries are developed. However, as the largest fields in any frontier area are usually found early in the exploration cycle, the increase will end and in the long term a general decline will once again set in.

Inferred oil reserves (projected field growth) are estimated to be 4,600 million barrels, as many new large fields in the region are yet to be developed. Table 7.23 contains estimates of undiscovered recoverable oil resources in the Rocky Mountain–northern Great Plains region by area. The high (5 percent chance) and low (95 percent chance) estimates for the region are 39,900 million barrels and 12,900 million barrels are not additive. The mean total (23,500 million barrels) is the expected volume and is additive. It can be seen from the table that the best prospects in the region for future oil discoveries are thought to be in the Overthrust Belt. The large area of southern Idaho, western Utah, and eastern Nevada (the eastern Basin and Range region) is also thought to have good oil potential. Similarly, future exploration drilling in the Paradox, Uinta, Piceance, Williston, Powder River, Green River, and Big Horn basins would appear to have good probabilities for oil discoveries.

The estimated ultimate oil recovery for the region, about 38.2 billion barrels, is shown in Table 7.24. The region appears to contain excellent oil prospects. In spite of the long history of exploration and the early discovery of many giant fields, an estimated 62 percent of the total projected recoverable oil remains to be found. However, 31 percent of this prospective oil is in the Overthrust Belt and will be much more expensive and difficult to find and produce than were the early large discoveries. Also, the new fields found in the areas that have been explored for many years will be smaller than the older fields and a large number will be needed to offset declines in older field production.

The past decade of natural gas production in the region is shown by state in Tables 7.25 through 7.30. In Wyoming, 1.703 trillion cubic feet more gas has been added to reserves than were produced in the past decade. Production has increased, along with the increase in reserves, and can be expected to continue to increase as overthrust and deep basin gas reservoirs are discovered and developed. In 1980, 0.416 trillion cubic feet of gas was produced from a reserve of 5.598 trillion cubic feet (see Table 7.25).

In Colorado, 0.722 trillion cubic feet more gas was added to reserves during the past decade than was produced. Both production and reserves increased with successful exploration for shallow gas in the Denver-Julesburg basin. The largest Colorado gas discoveries, however, were made in 1950. In 1980, 0.200 trillion cubic feet of gas was produced from a reserve of 2.299 trillion cubic feet (see Table 7.26).

North Dakota gas reserves increased at the end of the decade, as gas was recently discovered in the deeper parts of the Williston basin. Favorable reserves/production ratios (15:1) and new discoveries indicate that near-term North Dakota gas production may remain relatively stable or increase. During the past decade, however, 0.112 trillion cubic feet less gas was added to reserves than was produced. Production in 1980 was 0.037 trillion cubic feet from a reserve of 0.537 trillion cubic feet (see Table 7.27).

Montana gas production has increased and additional reserves have recently been discovered, especially in the Sweetgrass arch area. The high reserves/production ratio (21:1) indicates that gas production can be further increased as demand and pipeline construction allows. In the past decade, 0.177 trillion cubic feet more gas was discovered than was produced. In 1980, 0.061 trillion cubic feet was produced from a reserve of 1.287 trillion cubic feet (see Table 7.28).

Utah gas production increased at the end of the decade to 0.061 trillion cubic feet, but reserves fell during the decade by 21 percent, as 0.234 trillion cubic feet less gas was added to reserves than was produced. As Overthrust Belt fields are developed in Utah, gas production can be expected to continue to rise. Gas reserves in 1980 were 0.855 trillion cubic feet (see Table 7.29).

Although the largest fields were discovered in the 1920s and 1940s, western New Mexico gas production and reserves remain the most significant in the Rocky Mountain region. Gas production during the past decade was relatively stable and increasing reserves (0.833 trillion cubic feet more gas was added to reserves than was produced) have raised reserves/production ratios to 18:1. Prospects for continued stable production, at least in the near term, are good. Production in 1980 was 0.549 trillion cubic feet from a reserve of 9.908 trillion cubic feet (see Table 7.30).

The recent history of gas production in the region as a whole is shown in Table 7.31. For the past several years the region has experienced increasing gas production and reserves, as 3.089 trillion cubic feet more gas was added to reserves in the past decade than was produced. With reserves/production ratios relatively high (15:1), the prospects for stable or increasing gas production in the near term are good. The increase in reserves to 20.484 trillion cubic feet during the past several years has resulted from gas discoveries in the Overthrust Belt, the deep composite basins of Wyoming, the Denver-Julesburg basin, the deeper Williston basin, and the Sweetgrass arch of Montana, and from infill drilling in the San Juan basin of northwestern New Mexico. Production in 1980 increased to 1.324 trillion cubic feet. Inferred gas reserves (projected growth of discovered fields) are estimated to be 11.1 trillion cubic feet.

Table 7.32 contains estimates of undiscovered recoverable gas resources of the region. The high and low estimates are not additive, but they have been calculated for the region as 211.4 and 83.1 trillion cubic feet. The mean total (136.2 trillion cubic feet) is the expected volume for the region. The Potential Gas Committee recently assessed essentially the same region and projected undiscovered gas reserves at 151 trillion cubic feet. Of this total potential gas, 69 trillion cubic feet was estimated to occur below 15,000 feet, and 79 trillion cubic feet (more than half) was estimated to exist in the Overthrust Belt. Most of the undiscovered overthrust gas (56 percent) was estimated to occur below 15,000 feet.[46] The U.S. Geological Survey and the Potential Gas

Committee, then, have quite similar projections of undiscovered gas in the region despite differences in assessment procedure and personnel and the geological unknowns involved.

The Overthrust Belt is considered to have by far the greatest conventional gas potential of any area in the region. The Green River basin also is quite prospective. Other areas of promise are the eastern Basin and Range region and the Uinta and Piceance basins. The estimated ultimate gas recovery for the region is about 197.2 trillion cubic feet, as shown in Table 7.33. About 69 percent of the conventional gas in the region is estimated as yet to be discovered. However, just under half of this undiscovered gas is projected to occur below 15,000 feet, and significant amounts of the yet-to-be-discovered shallower gas is thought to be located in complex geological regions, such as the upper thrust plates of the Overthrust Belt. Thus, although the region contains excellent gas prospects, remaining fields will be more expensive and difficult to find and produce than were the gas fields discovered in the past.

The Energy Information Administration estimated 1980 natural gas liquids production at 53 million barrels for the Rocky Mountain–northern Great Plains region. Natural gas liquids reserves were estimated at 1,071 million barrels. Natural gas liquids production averaged about 40 barrels per million cubic feet of produced gas. From this average, the ultimate natural gas liquids recovery for the region, about 6.7 billion barrels, can be estimated and is shown in Table 7.34. The undiscovered resources were projected by the U.S. Geological Survey.

In summary, the petroleum remaining in the northern Great Plains and the Rocky Mountains is estimated at about 29.7 billion barrels of oil, 5.5 billion barrels of gas liquids, and 167.8 trillion cubic feet of gas, as shown in Table 7.35. When natural gas liquids are included, the natural gas deposits are estimated to contain more total energy than the oil deposits. Most of the remaining petroleum in the region is estimated to be still undiscovered.

West Texas and Eastern New Mexico

Petroleum Geology

A large, Paleozoic, intracontinental composite sedimentary basin occurs in the West Texas–eastern New Mexico region. The major portion of the larger region comprises the Permian basin, which is separated from the adjacent, smaller Fort Worth basin to the east by the Bend arch. To the south of the Permian basin, the Marathon fold belt occurs (see Figure 7.2).

Most of the Pennsylvanian period in the West Texas–eastern New Mexico region was characterized by expanding seas between emergent landmasses. The positive areas supplied clastic detritus to portions of the submerged areas. Carbonate rocks were deposited on the margins of the region, but organic limestone reefs were more widely distributed throughout the area. In Early Permian time, portions of the region were selectively uplifted and the deeper parts of the basins began to fill with fine clastic sediments while carbonates continued to form along the margins.

The region then subsided and large barrier-reef masses formed. The arid climate caused the reef-locked lagoons to become huge evaporative pans containing hypersaline water. Highly concentrated brines seeped slowly downward through the carbonate sediment of the lagoon floors and magnesium in the brines replaced the calcium in the carbonates, causing extensive dolomitization and extraordinary porosity. The sediments then were sealed by the deposition of evaporites, which became cap rocks. The source of the oil in the region was probably not Permian rocks, because evaporite deposits usually indicate conditions of depposition that are not favorable for prolific life. There are, however, deeper (older) fossiliferous formations that may have been the source beds. As the seas became progressively more restricted, the evaporite deposition encroached over the entire region. A relatively thin cover of Cretaceous, Tertiary, and Quaternary sediments was subsequently deposited.[47]

Petroleum was first discovered in the region in 1920. It occurs in anticlines, fault structures, stratigraphic traps, and reefs. The predominant reservoirs are carbonate rocks, although petroleum also occurs in sandstones and siltstones. Although Permian carbonates are by far the most important reservoirs, production also is derived from Cambrian- to Cretaceous-age rocks. In the Bend arch, the most productive reservoir rock is of Early Pennsylvanian age.

The region contains 33 national-class or larger giant oil fields. The largest of these is the Yates field, a world-class giant, discovered in 1926. It is in a fold cut by faults, and its major reservoir is a Permian carbonate. Another world-class giant field, discovered in 1936, is Wasson, a dolomitized Permian reef accumulation affected by local folding.[48] Recoverable Wasson reserves were recently significantly increased due to the success of a major waterflood project and considerable infill drilling. An enhanced recovery project utilizing carbon dioxide is also being considered for the field.[49]

Recent exploratory drilling has shifted from the shallower oil prospects to the deeper gas trends. Several deep high-potential gas discoveries have been made near the New Mexico state line, including Pennsylvanian gas below 20,000 feet. Silurian-Devonian structural and stratigraphic traps have also been drilled. Development drilling continues at a high rate as the area matures as a petroleum-producing region. Injection wells for waterfloods are also being drilled. The development of New Mexico's carbon dioxide reserves will allow better exploitation of the Permian basin oil accumulations, as carbon dioxide will be used for enhanced oil recovery operations in the region throughout the 1980s.[50]

Production, Reserves, and Resources

A historical sequence of the region's oil production is shown in Tables 7.36, 7.37, and 7.38. West Texas reserves have declined by 25 percent (to 4,980 million barrels in 1980) in the past decade, while production has declined by only 4 percent (to 607 million barrels). Thus, production has been maintained by a drawdown of reserves. Reserves/production ratios have fallen from 11:1 to 8:1, as 1,681 million more barrels of oil were produced than were added to reserves (see Table 7.36). The recent increase in reserves is attributable to upward revisions in such older national- and world-class giant fields as Levelland, Seminole, Salt Creek, Slaughter, and Wasson, rather than to new field discoveries. The revisions reflect the results of waterflood and in-fill drilling projects.

During the past decade eastern New Mexico reserves declined about 47 percent (to 427 million barrels in 1980) as 384 million more barrels of oil were produced than added to reserves. Reserves/production ratios were already low at the beginning of the decade, meaning that production also had to decline (along with declining reserves); it fell by 46 percent to 63 million barrels in 1980 (see Table 7.37). The most important oil discovery in eastern New Mexico was made in 1929, when the Vacuum field, a national class giant, was discovered. It is an anticlinal structure with a Permian reservoir.

Oil production in the region is down 10 percent, and reserves have declined by 28 percent as 2,065 million more barrels were produced than were added to reserves. Reserves/production ratios have also declined. The region is a mature petroleum province in which recent reserve additions have mostly come from the upward reserve revisions resulting from waterflooding and in-fill drilling of older, large fields. Oil production in 1980 was 670 million barrels from a reserve of 5,407 million barrels (see Table 7.38).

Inferred oil reserves (projected growth of discovered fields) are estimated to be 5,300 million barrels. Table 7.39 contains estimates of undiscovered recoverable oil resources in the region. It can be seen from the table that the best prospects are considered still to be in the Permian basin, in which exploration drilling has recently increased.

Exploratory activity also continues in the Marathon fold belt. The low and high undiscovered recoverable oil resources estimates for the entire region have been calculated as 2,700 and 9,400 million barrels. The expected volume is 5,400 million barrels.

The estimated ultimate oil recovery for the region (about 42 billion barrels) is shown in Table 7.40. According to the table, 87 percent of the oil estimated to be ultimately recoverable from the region has already been discovered. Although there is a reasonably large resource remaining to be found, it will occur in smaller and harder to find fields, making exploration more difficult and expensive. The area has experienced a significant decline in discovery rate (the amount of petroleum found per foot of exploratory drilling).

The discovery pattern in the Permian basin has been found to fall into three phases. In the initial phase, from 1920 to 1938, the discovery rate was high and 2,015 exploratory holes found 31 of the 70 largest fields discovered to date in the basin. During the second phase, from 1939 to 1950, 4,271 exploratory holes discovered 24 additional large fields. This was a transitional phase during which the volume of petroleum discovered per foot of exploratory drilling fell rapidly. The third phase began in 1951, and since then only 15 of the 70 largest fields have been discovered by over 25,000 exploratory holes. The discovery rate has remained low and stable, with numerous but small new fields found. This trend is expected to continue.[51]

The past decade of natural gas production from the region is shown in Tables 7.41, 7.42, and 7.43. Natural gas production in West Texas declined by 12 percent in the past decade; but proved gas reserves fell by 60 percent, as 16.966 trillion cubic feet more gas was produced than was added to reserves. Production during the decade included the steady drawdown of reserves, resulting in a reserves/production ratio in 1980 of 6 : 1. This low ratio indicates that many fields are nearing depletion and that future increases in gas production will have to come from new gas discoveries rather than from a further drawing down of reserves. A recent increase in reserve additions was primarily from reserve revisions in existing fields rather than from new field discoveries. Gas production in 1980 was 1.775 trillion cubic feet from a reserve of 11.187 trillion cubic feet (see Table 7.41).

Natural gas production in eastern New Mexico was held to a 4 percent decline during the past decade by a 39 percent drawdown of proved reserves to 3.150 trillion cubic feet in 1980, as 2.047 trillion cubic feet more gas was produced than was added to reserves. Reserves/ production ratios fell to 6:1. Gas production in 1980 was 0.515 trillion cubic feet (see Table 7.42).

In the region as a whole, gas production fell 10 percent (to 2.290

trillion cubic feet in 1980), but reserves declined by 57 percent, as 19.013 trillion cubic feet more gas was produced than was added to reserves during the decade. Therefore, gas has been produced mostly by the drawing down of reserves. With a reserves/production ratio of 6:1, future production will depend primarily on new field discoveries. Such discoveries will have to be made in spite of declining discovery rates or gas production will fall at an accelerating pace. Gas reserves in 1980 were 14.337 trillion cubic feet (see Table 7.43).

Inferred gas reserves (projected field growth) for the region are estimated to be 16.3 trillion cubic feet. Table 7.44 contains estimates of undiscovered recoverable gas resources. The expected quantity of undiscovered gas is 42.7 trillion cubic feet. The best gas prospects are considered to be in the Permian basin, where deep gas exploration drilling is under way. The Potential Gas Committee estimates that a total of 35 trillion cubic feet of gas remains to be found in the Permian basin, 15 trillion cubic feet of which is below 15,000 feet. The high and low estimates for the region are not additive, but have been calculated to be 75.2 and 22.4 trillion cubic feet.

The estimated ultimate conventional gas recovery for the region (about 147.7 trillion cubic feet) is shown in Table 7.45. About 71 percent of the total recoverable gas estimated to exist in the region has already been discovered, and much of the undiscovered gas is thought to occur below 15,000 feet. Thus, the remaining gas will be found in deeper or in smaller, shallower fields than in the past. Exploration will be slower and more expensive. Gas production, therefore, may continue to decline in the region, although there are large inferred reserves that will help future production.

The Energy Information Administration estimated 1980 gas liquids production in the region at 172 million barrels from a proved reserve of 1,309 million barrels. Average gas liquids production was, therefore, about 75 barrels per million cubic feet of produced gas. From this average, the ultimate natural gas liquids recovery for the region (about 9.6 billion barrels) can be estimated and is shown in Table 7.46. Undiscovered resources were projected by the U.S. Geological Survey.

In summary, the estimates of petroleum remaining in the West Texas and eastern New Mexico region (about 16.1 billion barrels of oil, 4 billion barrels of gas liquids, and 73.3 trillion cubic feet of natural gas) are shown in Table 7.47. Natural gas production may continue to decline, but this decline may be moderated somewhat by the development of inferred reserves. Oil production could possibly be stabilized in the short term by in-fill drilling and waterflooding, but as this is unlikely, the oil production decline is expected to continue.

Gulf Coast

Petroleum Geology

The Gulf Coast region includes the Texas Gulf area, Louisiana, Mississippi, Alabama, Arkansas, northwestern Florida, and offshore Gulf of Mexico. The Gulf Coast, or Western Gulf, basin is an open extracontinental downwarped basin (with a superimposed delta basin) with a depositional history dating from the Jurassic period, or possibly even earlier. Throughout Early and Middle Jurassic time, much of the Gulf region was covered by shallow seas in which restricted circulation permitted the extensive precipitation of evaporites such as salt and anhydrite. By Early Cretaceous time, active organic reef complexes had formed and extended over a large portion of the southern and eastern Gulf area. Reef growth continued through the Cretaceous and Tertiary periods, accompanied by regional subsidence. At the end of the Mesozoic era, a period of widespread uplift elevated the Rocky Mountains and the High Plains, and subsequent erosion of this terrain during Tertiary time supplied vast quantities of clastic sediments (sands, silts, and clays) to the rivers draining into the north-central and northwestern Gulf. These clastic sediments accumulated in the slowly subsiding Gulf Coast basin. Sedimentation continued throughout the Cenozoic era and resulted in the southward progradation of the Gulf coastal plain. However, the basin regions receiving maximum deposition migrated northeastward to the present location of the Mississippi Delta.

The Gulf of Mexico continental margin consists of an eastern geological province (offshore Florida) composed of shallow carbonate banks and relatively simple geological structures and a western geological province of clastics with complex structure involving salt mobilization. Much of the lithological and structural character of the western Gulf is related to the thick clastic sedimentary sequence that accumulated during Cenozoic time and to the extensive Jurassic evaporite deposits that underlie this sequence. As the subsiding Gulf Coast basin filled with compacting sediments, in the course of geologic time petroleum was generated from the contained organic matter and migrated into the sandstone formations.

The most important tectonic features of the region are the numerous large salt masses that have moved upward through the sediments in response to the vertical pressure of the accumulating overburden and the density differences between the lighter salt and the heavier clastic materials. Many of the petroleum traps are structures that are related to the intruded salt masses (see Figure 1.3), but petroleum accumulations also occur in other fold and normal fault traps and in stratigraphic traps. The most productive sequence in the area has been the Miocene sediments of Louisiana, but the Pliocene and Pleistocene series also contain commercial reservoirs. In the Louisiana Gulf region, the reservoirs become younger from north to south, having been deposited as the coastline regressed southward. There are a number of world-class giant fields in the Louisiana portion of the region. The Tambalier Bay and Caillou Island fields in southeast Louisiana are part of a salt-dome complex that penetrates to within 2,000 to 3,000 feet of the surface, forming a variety of traps in Miocene- and Pliocene-age sandstones. Other Louisiana world-class giant fields include Bastian Bay, Bateman Lake, and Bayou Sale.

World-class giant fields also occur in the Gulf Coast part of Texas, where petroleum accumulations of varying sizes are found in anticlines, fault structures, unconformity traps, and other stratigraphic traps but most frequently occur over and around salt domes. Production is from Eocene, Oligocene, and Miocene reservoirs. The world-class giants include Greta–Tom O'Connor, Borregos-Seeligston, and Agua Dulce-Stratton, which are large anticlinal accumulations along an early Oligocene producing trend that parallels the shoreline. Conroe, discovered in 1931, is another world-class giant field. It is in a broad, gentle structure that is probably related to deep salt movement. Production is from Eocene reservoirs. Hastings is a faulted anticlinal oil accumulation overlying a buried salt mass. Other Texas Gulf world-class giant petroleum fields are: Van, West Ranch, Katy, Old Ocean, Webster, Thompson, Pledger, and La Gloria.

There was about a 35 percent drilling increase in the upper Texas Gulf Coast in 1980 compared to the previous year. Onshore development drilling accounted for most of this increase, but onshore exploratory drilling also increased. The Oligocene trend was the most actively explored onshore play. Offshore exploratory drilling also increased. The onshore Louisiana Gulf Coast continued to be one of the most intensely explored regions of the Gulf Coast basin. Most of the new discoveries were in the prolific Miocene trend.[52] Offshore Louisiana drilling decreased slightly from the previous year.

Further north in the Louisiana portion of the Gulf Coast basin, the Late Cretaceous Tuscaloosa sandstones have recently been found to produce natural gas and condensate along a 20- to 30-mile-wide belt that extends from northeast of New Orleans through Baton Rouge to the Texas border. The Tuscaloosa sands were deposited in shallow water over an Early Cretaceous shelf-edge reef complex. Large amounts of clastic material were deposited as deltas and submarine fans as streams breached the reef edge. Parallel to and about forty miles south of the Early Cretaceous shelf edge is a system of regional growth faults downthrown to the southwest. This fault system divides the Tuscaloosa trend into two portions, a normally pressured zone to the north and a geopressured zone to the south. Depth, temperature, and pressure generally increase to the south, toward the deeper part of the basin. Most of the traps in the Tuscaloosa trend are controlled by structure (the fault system), but stratigraphic traps also exist. Exploration is conducted by seismic surveys and there is excellent resolution and definition at depths of up to 25,000 feet.[53]

Thus far, wells have been drilled to between 16,000 and 26,000 feet. The drilling is difficult with temperatures up to 425 degrees Fahrenheit, pressures up to 16,000 pounds per square inch, and carbon dioxide and hydrogen sulfide occurrences. Since the Tuscaloosa play began in 1975, there has been considerable drilling activity along the trend, and about 9 fields have been discovered.[54] The exploration success level, however, has dropped from nearly 100 percent to 20 percent in 1980, and most of the discoveries to date have been concentrated in a central structural complex. The ultimate gas potential of this trend appears significant, perhaps about 11 trillion cubic feet.[55]

The Sabine uplift is a broad domed subsurface structure about 80 miles in diameter located in northeastern Texas and northwestern Louisiana. It is the dominating structural influence on petroleum accumulation in the area. Several important fields in eastern Texas, southern Arkansas, and northern Louisiana are associated with this uplift.

The East Texas field, one of only two super-giant fields in the United States, lies on the western flank of the Sabine uplift in the East Texas basin (which is part of the Gulf province). Discovered in 1930, its productive area covers about 140,000 acres and originally held about 6 billion barrels of recoverable oil. It is a large stratigraphic trap, with Late Cretaceous Woodbine shoreline sands wedging out in a facies change and unconformably overlying Early Cretaceous strata. Over 11,600 wells have been drilled in the East Texas field. Hawkins, discovered in 1940, is a world-class giant field in the eastern Texas area. Drilling in the East Texas basin has more than doubled over the past three years.

The northern Louisiana salt basin region produces from rocks of Jurassic, Cretaceous, and Tertiary age. The petroleum occurs in Jurassic and Cretaceous sandstone and carbonate reservoirs in anticlinal traps associated with salt tectonics and also in fault traps and stratigraphic traps. Tertiary production is primarily from reservoirs associated with a deltaic environment.[56] Many of the discoveries in the area have been on the flank of the Sabine uplift, the dominant local tectonic feature. The largest petroleum accumulation in northern Louisiana is the Monroe gas field, a world-class giant covering an area of about 425 square miles. Discovered in 1916, the accumulation is primarily a stratigraphic trap on a regional uplift. Gas is produced from two Cretaceous reservoirs. The largest oil field was discovered in 1905. Caddo-Pine Island is a national-class giant discovered on a large, low anticline by random drilling. Exploration drilling in the area has increased by about 45 percent over the past decade.

The salt basin of Mississippi, Alabama, and northwestern Florida is also productive and an area of increased exploration. The leading drilling region is in southwestern Mississippi where the primary reservoirs are shallow Eocene clastics. The basin also contains sandstone and carbonate reservoirs of Jurassic and Cretaceous age. The Jay field in northwestern Florida is a national-class giant oil field, producing from the Jurassic Smackover limestone and dolomite. Gas and minor amounts of oil are being produced from Paleozoic rocks, mainly Mississippian-age sandstones, in the Black Warrior basin of northern Mississippi and northwestern Alabama. However, none of the recent discoveries was considered significant and extensive development programs are unlikely.[57]

In the Gulf coastal plain of Arkansas, anticlinal, fault, and stratigraphic traps occur that contain petroleum in Jurassic and Cretaceous reservoirs. The major field is the Smackover oil field, a world-class giant discovered in 1922. The Arkoma basin, in the northwestern part of the state, produces nonassociated gas from several Paleozoic formations. (This basin extends into Oklahoma and will be discussed in the Midcontinent section.)

Production, Reserves, and Resources

A state-by-state historical sequence of the region's oil situation is shown in Tables 7.48 through 7.54. Between 1969 and 1980, 3,741 million more barrels of oil were produced than discovered in the Texas Gulf Coast area, and reserves declined by 61 percent. Production fell by 29 percent, partially supported by the drawing down of reserves. As the reserves/production ratio has reached 8:1, production will have to depend more upon reserve additions in the future. Production in 1980 was 315 million barrels from a reserve of 2,425 million barrels (see Table 7.48).

During the decade, 3,161 million more barrels of oil were produced than added to reserves in Louisiana, causing reserves to decline by 56 percent. As the reserves/production ratios were relatively low, however, even at the beginning of the decade, it was not possible to slow steep production declines by more rapidly drawing down reserves: as reserves fell, production fell almost as rapidly (45 percent). The 6:1 reserves/production ratio represents a near-maximum annual withdrawal rate, so it will take a significant increase in reserve additions to arrest the production decline. Production in 1980 was 400 million barrels from a reserve of 2,528 million barrels (see Table 7.49).

In Mississippi, reserves declined by 60 percent during the past decade, as 217 million more barrels of oil were produced than were found. As reserves/production ratios remained low, production dropped by 53 percent, and as increased production depends on increased reserve additions, it will be difficult to arrest the production decline in this mature area. Production in 1980 was 30 million barrels from a reserve of 143 million barrels (see Table 7.50).

During the past decade reserves declined by 49 percent in Alabama, as 33 million barrels less oil were added to reserves than produced. Production, however, actually increased as reserves were rapidly drawn down. Reserves/production ratios of 3:1 and 4:1 indicate fields that are near depletion in very maturely developed areas. Future production will depend upon increased reserve additions. The 1980 reserves did increase. This trend will have to be continued or production will fall. Production in 1980 was 9 million barrels from a reserve of 34 million barrels (see Table 7.51).

Twenty million more barrels of oil were produced in Arkansas than were added to reserves during the past decade. Reserves were drawn down about 16 percent, but production and reserves/production ratios remained about the same. In 1980, 16 million barrels of oil were produced from a reserve of 107 million barrels (see Table 7.52).

During the decade Florida oil production increased by 21 times to 42 million barrels while reserves increased by a factor of 14.3, to 129 million barrels, as 120 million barrels more oil were added to reserves than were produced (see Table 7.53). Most of this oil was in the national-class giant Jay field, discovered in 1970. Although it provides about half of Florida's oil production, the field is now 75 percent depleted, and Florida reserves/production ratios are again very low. Production will continue the drop that began in 1980 unless significant new oil discoveries are made.

Total Gulf Coast oil production history is shown in Table 7.54. During the decade 7,052 million barrels more oil were produced than added to reserves in the region; thus, reserves declined by 57 percent. Production fell by 36 percent and reserves/production ratios also declined. To arrest the decline of production more reserves will have to be discovered. Reserve additions were reasonably good in 1980 (648 million barrels), but remained lower than production. Gulf region production in 1980 was 812 million barrels from a reserve of 5,366 million barrels.

Inferred oil reserves for the region are estimated to be 6,300 million barrels. Table 7.55 contains estimates of undiscovered recoverable oil resources in the region. The expected quantity of 13,900 million barrels is additive. It can be seen from the table that the best prospects for reserve additions remain in the Western Gulf basin. The Louisiana and Mississippi salt basins also are quite prospective, as is the Eastern Gulf basin. The high and low undiscovered oil resource estimates are not additive, but have been calculated for the region as a whole as about 13,100 and 3,600 million barrels onshore and 11,100 and 3,100 million barrels offshore.

The estimated ultimate recovery for the region is about 66.9 billion barrels, as shown in Table 7.56. According to the table, 79 percent of the oil estimated to be ultimately recoverable from the region has already been discovered, and much more oil has been produced than remains to be found. A significant undiscovered resource is thought to remain, but it will be mostly in numerous small (and more expensive to find) fields rather than in the larger fields of the past.

The past decade of natural gas production in the region is shown in Tables 7.57 through 7.62. Texas Gulf Coast gas reserves have declined by 59 percent, as 44.032 trillion cubic feet more gas was produced than was added to reserves. Due to revisions, reserve additions are statistically negative. Production has decreased only by 15 percent due to the drawing down of reserves. Reserves/production ratios have fallen to a level that requires future production to be more directly related to discoveries; more gas will have to be found if production is to be maintained. As shown in Table 7.57, production in 1980 was 3.863 trillion cubic feet from a reserve of 30.888 trillion cubic feet.

In spite of the discovery of over 47 trillion cubic feet of gas in Louisiana during the past decade, 42.163 trillion cubic feet more gas was produced than was added to reserves. Thus, reserves declined by almost 50 percent. Production, however, was down by only 7 percent, due to the drawdown of reserves. Reserves/production ratios were halved during the decade and are now 6:1, indicating that future production will depend upon reserve additions. More gas will have to be added to reserves in the next decade than was added in the last if current gas production is to be maintained. According to Table 7.58, production in 1980 was 6.744 trillion cubic feet from a reserve of 42.781 trillion cubic feet.

Both production and reserves in Mississippi increased during the decade (2 and 12 percent, respectively); more gas was found than was produced. Even so, the state is a relatively small producer of natural gas: in 1980, only 0.170 trillion cubic feet of gas was produced from a reserve of 1.582 trillion cubic feet (see Table 7.59).

Significant gas discoveries were made in Alabama during the last decade and gas production increased to 0.064 trillion cubic feet in 1980 (see Table 7.60). Reserve additions during this time were 0.853 trillion cubic feet in excess of production. Reserves (0.910 trillion cubic feet) can be drawn down in the future to help support production. Alabama gas production, although expanding, is quite small in comparison with other Gulf Coast states.

During the decade Arkansas gas production declined by 30 percent and reserves declined 36 percent, as 0.923 trillion cubic feet more gas was produced than was added to reserves. Reserves/production ratios remained relatively high (14:1); thus, reserves could be drawn down in the short term to support production. Reserve additions in 1980 exceeded production, as reflected in an increased reserves/production ratio. Production in 1980 was 0.117 trillion cubic feet from a reserve of 1.673 trillion cubic feet (see Table 7.61).

The total Gulf Coast gas producing history of the past decade is shown in Table 7.62. Gas reserves in the region declined by 53 percent during the past decade, as 86.089 trillion cubic feet more gas was produced than was added to reserves. This drawdown of reserves supported production to the extent that the latter declined by only 10 percent. Reserves/production ratios fell by one-half to 7:1, indicating that reserves drawdown cannot continue to support production to the same extent as in the past. Future production will have to depend more on reserve additions, which increased somewhat in 1980 but were still below production levels. If these trends continue, Gulf Coast gas production will continue to fall. Production in 1980 was 10.958 trillion cubic feet from a reserve of 77.834 trillion cubic feet.

Inferred gas reserves (projected discovered field growth) for the region are estimated to be 111.1 trillion cubic feet. The conversion of these inferred reserves to proved reserves will be an important factor in sustaining future Gulf Coast gas production.

Table 7.63 contains estimates of undiscovered recoverable gas resources in the region (with an expected quantity of 197.7 trillion cubic feet). As is the case with oil, the best prospects for gas discoveries remain in the Western Gulf basin. The low and high undiscovered gas resource estimates have been calculated for the region at about 56.9 and 252.5 trillion cubic feet onshore and 41.7 and 114.2 trillion cubic feet offshore.

The estimated ultimate Gulf Coast gas recovery is shown in Table 7.64 as about 682.7 trillion cubic feet. It can be seen from the table that considerably more gas has already been produced than is estimated to remain. About 71 percent of the total gas thought to exist in the Gulf Coast region has already been discovered.

The Energy Information Administration estimated 1980 gas liquids production in the region at 320 million barrels from a proved reserve of 2,591 million barrels. Average gas liquids production was therefore 29.2 barrels per million cubic feet of produced gas. From this average, the ultimate natural gas liquids recovery from the region (about 20.0 billion barrels) is estimated in Table 7.65. Undiscovered gas liquids resources were projected by the U.S. Geological Survey as 5,600 million barrels.

The total remaining recoverable petroleum estimated to occur in the region is shown in Table 7.66 as about 25.6 billion barrels of oil, 11.4 billion barrels of gas liquids, and 386.6 trillion cubic feet of gas. There is more energy remaining in gas deposits than in oil deposits. Both oil and gas reserves/production ratios are relatively low, indicating that future production will have to depend more upon reserve additions than upon reserve drawdown as in the past decade. Substantial petroleum resources exist, but new discoveries can be expected to be smaller and deeper and thus more expensive to produce. Oil production will probably continue its decline, but gas production may stabilize in the near term as substantial inferred reserves and undiscovered resources become available for conversion to proved reserves.

Midcontinent

Petroleum Geology

The Midcontinent region includes the Texas Panhandle, Oklahoma, Kansas, and Missouri. The major basin in the region is the Anadarko, a large intracontinental composite basin located in western and central Oklahoma, the Texas Panhandle, and southern and western Kansas (see Figure 7.2). This basin contains more than 40,000 feet of Paleozoic sediments, many horizons of which have produced petroleum. In the past, most of the basin's output has come from its vast shelves, including the Hugoton and Sedgwick embayments of Kansas, but more recently the deep axial trough of the basin also has been tested. Hundreds of petroleum fields have been discovered in the basin, in both structural and stratigraphic traps.

The Anadarko basin may be divided into five major geologic time sequences. The first extends from the Middle Cambrian to a post-Devonian orogeny. Reservoir formations in this sequence (Arbuckle, Simpson, Viola, and Hunton) are almost entirely carbonates. The second sequence is Mississippian in age and is mostly carbonates. The Mississippian gas reserves in the basin occur primarily in stratigraphic traps. The third sequence consists of Pennsylvania Morrow-Springer clastics. Petroleum reserves in this sequence also are in stratigraphically controlled traps. It is one of the most important producers in the basin and ranks second in petroleum reserves. The fourth sequence is Late Pennsylvanian. Its deposits are mostly clastics, and its reserves are primarily in stratigraphic traps. The last and most prolific sequence is the Permian, which contains the world-class giant Hugoton-Panhandle gas and oil field.

There are four major productive trends in the Anadarko basin. The first is on the northern rim where petroleum is trapped in the subcrop of the Hunton and in isolated erosional remnants. The second trend is along the eastern edge of the basin where movement on the buried Nemaha granite ridge has caused considerable faulting (creating structural traps) in early Paleozoic sediments. The third trend is along the hinge line of the basin, north of the deep basinal axis, where faulting has created a linear pattern of structural traps. The fourth trend is adjacent to the Amarillo-Wichita uplift, which caused 10,000 to 20,000 feet vertical displacement in early Paleozoic rocks during post-Morrowan time.[58]

The deep Anadarko gas potential was first tested in 1969 with a 21,600-foot hole in the Springer formation. Since that time, and particularly since the Natural Gas Policy Act of 1978 removed federal price controls from gas occurring below 15,000 feet (on November 1, 1979), deep drilling has become common, with several hundred wells drilled below 15,000 feet, including two below 30,000 feet. The deep Anadarko basin is thought to contain about 3 trillion cubic feet of gas reserves, with an undiscovered potential ranging from 9 to 39 trillion cubic feet. The higher estimated recoveries, however, may be very difficult to achieve because the sediments below 20,000 feet constitute only about 10 percent of the total volume of the basin's contained sediments and there are significant drilling problems at these depths. As in most frontier areas, the best prospects may be expected to be tested early in the exploration cycle and the largest accumulations discovered relatively soon. Subsequent deep drilling, therefore, will be testing increasingly poorer prospects. Deep drilling decreased in 1980, although the success rate was excellent (67 percent of the wells were completed as potential producers).[59] The cost of such deep wells ($3 to $10 million) will prevent the drilling of numerous marginal prospects, as is often done at shallower depths.

Glenn Pool was the first large field to be found in Oklahoma. Discovered in 1906, its production peaked in 1908 and then steadily declined. Burbank, discovered in 1920 in northern Oklahoma, is a world-class giant field. It is a stratigraphic trap affected by local structure. Accumulation has been controlled by variations in the porosity of a Pennsylvanian sandstone that wedges out updip. Another world-class giant is Oklahoma City, which was discovered in 1928 and is now in an advanced stage of decline. The field is in a faulted anticline with oil production from Pennsylvanian and Simpson (Ordovician) sandstones. The major reservoir is the Cambrian-Ordovician Arbuckle limestone. In southern Oklahoma, the national-class giant Hewitt, discovered in 1919, is an anticline overlying a buried uplift. Oil is derived from several Pennsylvanian sandstones. Healdton, also a national-class giant, is similar in structure, with production from Pennsylvanian sandstones and Arbuckle limestones.

Drilling activity in Oklahoma and the Texas Panhandle, already at a high level, increased in 1980. Exploratory drilling increased by only 2 percent, but development drilling increased by 43 percent, as increasing prices encouraged in-fill drilling and more marginal field completions.

A shelf of the Anadarko basin (the Hugoton embayment bounded by the Las Animas arch) extends into Kansas. The sediments of the shelf are predominantly Paleozoic-age carbonates. The Hugoton embayment and Las Animas arch are petroleum productive, with the Permian limestones as especially good gas producers. The largest positive feature in Kansas is the Central Kansas uplift, which was eroded during Late Mississippian–Pennsylvanian time. During this erosional interval, numerous topographic features and stratigraphic conditions favorable to trapping petroleum were created. Later, Middle Pennsylvanian sediments covered the area. The Arbuckle sequence has produced most of the oil in the Central Kansas uplift area.[60]

The Paleozoic rocks of southeastern Kansas (in the Sedgwick embayment of the Anadarko basin and the Cherokee platform) have also been highly productive. Petroleum in this area occurs mostly in Pennsylvanian and Mississippian reservoirs. The basin and platform are separated by the Nemaha ridge. Large petroleum fields have been found in anticlinal structures on the flanks of this ridge. In eastern Kansas, paleogeomorphic traps in the form of long, narrow, Pennsylvania-age sand bodies occur and are known as "shoestrings" because of their sinuous shape. The Cherokee platform is also productive in Missouri.

To the north, the Forest City basin is located mostly in Iowa, Missouri, and Nebraska, with the extreme southwest corner in Kansas. Its maximum sediment thickness is only about 4,000 feet and it contains rocks of Paleozoic age. The Forest City basin is not an important producer, as it appears to lack cap rocks. It does, however, contain minor petroleum accumulations in both Kansas and Missouri.

One of the oldest and largest fields in Kansas is El Dorado, an national-class giant discovered in 1917. It is an anticlinal field in the Nemaha ridge area. The major limestone reservoirs are Ordovician in age, but oil and gas are also found in Pennsylvanian sandstones and

Mississippian limestones. Other national-class giants found in anticlinal structures on the flanks of the Nemaha ridge are: Bemis-Shutts, discovered in 1928; Chase-Silica, discovered in 1931; and Trapp, discovered in 1937. The Hugoton in southwestern Kansas is a world-class giant gas field. Over 7,000 wells have been drilled in this extensive field, which produces from a series of Permian limestones and dolomites. The gas accumulations are stratigraphically controlled by variations in lithology: the proportion of clastics increases westward, with a consequent reduction in reservoir porosity. The productive area extends along a 250-mile trend that includes the giant Panhandle gas and oil field in the Texas Panhandle.

Drilling increased in Kansas in 1980 by almost 20 percent over 1979. This drilling included some exploration in such unproved basins as Salina and Forest City. New discoveries, however, were primarily on the Anadarko shelves and the Central Kansas uplift, and there were also discoveries on the Nemaha ridge. Twenty-one percent of the total wells drilled in 1980 were for exploration, but there was only a 1 percent increase in exploratory drilling over 1979.[61]

The Arkoma basin of Oklahoma and Arkansas is a Paleozoic intracontinental composite basin that produces nonassociated gas primarily from Early Pennsylvanian sandstones. Other productive gas reservoirs are of Silurian, Devonian, and Mississippian age. Successful recent exploration has included deeper drilling in Cambrian and Ordovician rocks in the eastern portions of the basin. The largest gas reserves discovered so far in the Arkoma basin have been in Oklahoma.

Production, Reserves, and Resources

A historical sequence of the region's oil situation is shown in Tables 7.67 through 7.70.

Texas Panhandle oil production dropped by 48 percent during the past decade, while reserves declined by 41 percent, with 96 million more barrels produced than added to reserves. In 1980, 15 million barrels of oil were produced in the Texas Panhandle from a reserve of 139 million barrels (see Table 7.67).

Oklahoma oil production fell by 39 percent and reserves declined by 34 percent, as 468 million more barrels were produced than added to reserves during the past decade. Production in 1980 was 131 million barrels and reserves were 922 million barrels (see Table 7.68).

Kansas oil production dropped by 39 percent and reserves declined by 45 percent during the decade, as 256 million more barrels of oil were produced than added to reserves. Production in 1980 was 54 million barrels from 310 million barrels of reserves (see Table 7.69).

In the region as a whole, oil production fell by 40 percent in the past decade. Reserves declined by 37 percent, as 820 million more barrels of oil were produced than were added to reserves. With the reserves/production ratio relatively low (7:1), production will continue to depend primarily upon reserve additions and will continue to trend downward if reserve additions remain below production levels. Production in 1980 was 200 million barrels from reserves of 1,371 million barrels (see Table 7.70).

Inferred oil reserves for the region are estimated to be 1,500 million barrels. Table 7.71 contains estimates of undiscovered recoverable oil resources in the region. The best prospects for future Midcontinent oil discoveries appear to be in the Anadarko basin, and the expected quantity of undiscovered oil is 4,200 million barrels. The low and high undiscovered resources estimates are not additive, but have been calculated for the region as a whole by the U.S. Geological Survey as 2,300 and 7,700 million barrels, respectively.

Table 7.72 indicates that the estimated ultimate oil recovery for the region is about 25.6 billion barrels. Of the total oil estimated to exist, 84 percent has already been discovered. Nonetheless, there is certainly a sufficient resource to justify continued exploration, but perhaps not enough to ensure long-term production stabilization. The newly discovered fields will be smaller and more difficult to find and more expensive to produce than were the giants of the past.

The last decade of natural gas production in the Midcontinent region is shown in Tables 7.73, 7.74, and 7.75. Texas Panhandle gas production fell 15 percent, to 0.953 trillion cubic feet, during the period, but reserves declined by 37 percent (to 6.777 trillion cubic feet), as 3.954 trillion cubic feet more gas was produced than was added to reserves. Reserves/production ratios have fallen to 7:1 as reserves have been drawn down to support production (see Table 7.73). As reserves/ production ratios are now relatively low, it will not be possible to continue to support production from reserve drawdown to the same degree as in the past, and production will decline at a faster rate unless reserve additions are increased.

Oklahoma gas production during the past decade dropped by 9 percent to 1.526 trillion cubic feet and gas reserves declined by 40 percent, as 6.993 trillion cubic feet more gas was produced than was added to reserves. Gas production was maintained at the expense of reserve drawdown. As reserves/production ratios have decreased to 7:1, this will become more difficult in the future. Oklahoma gas reserves in 1980 were 10.393 trillion cubic feet (see Table 7.74).

Although 2.523 trillion cubic feet more gas was produced in Kansas during the decade than was added to reserves (and reserves thus declined by 18 percent), production dropped by 23 percent, causing the reserves/

production ratio to increase to 17:1. With the reserves/production ratio relatively high, reserve drawdown could help support production in the near term, which may result in the stabilization of a declining gas production. Production in 1980 was 0.682 trillion cubic feet from a proved reserve of 11.510 trillion cubic feet (see Table 7.75).

Total Midcontinent gas production history for the past decade is shown in Table 7.76. Gas reserves declined by 32 percent to 28.680 trillion cubic feet during the past decade, as 13.470 trillion cubic feet more gas was produced than was added to reserves. Gas production, however, declined by only 14 percent (to 3.161 trillion cubic feet), being partially supported by reserve drawdown. Such support may continue if reserves/production ratios continue to fall, but to a lesser extent than was possible during the 1970s. Thus, gas production will depend more upon reserve additions in the next decade than in the last.

Inferred gas reserves for the region are estimated to be 16.9 trillion cubic feet. Table 7.77 contains estimates of undiscovered recoverable gas resources in the Midcontinent region. The expected volume of undiscovered gas is 44.4 trillion cubic feet. It can be seen from the table that the best gas prospects are in the Anadarko basin. The low and high estimates for the region are not additive, but have been calculated as 22.9 and 80.8 trillion cubic feet, respectively, by the U.S. Geological Survey. The Potential Gas Committee is somewhat more optimistic, with a projection of 108 trillion cubic feet of undiscovered gas supply, 42 trillion cubic feet of which is considered speculative. Included in this projection are 76 trillion cubic feet of potential deep gas (39 trillion cubic feet listed as possible and 37 trillion cubic considered speculative).[62]

The estimated ultimate gas recovery from the region (about 218.7 trillion cubic feet) is shown in Table 7.78. Eighty percent of the recoverable conventional natural gas estimated to exist in the Midcontinent region has already been discovered. However, a substantial amount remains for exploration, particularly in the deep Anadarko basin, but this gas will be very expensive to find and produce.

Midcontinent natural gas liquids production in 1980 was 121 million barrels from a reserve of 1,067 million barrels.[63] There were 38.3 barrels of natural gas liquids produced per million cubic feet of produced natural gas. On that basis, the ultimate recovery of natural gas liquids in the region is calculated as about 7.9 billion barrels, as shown in Table 7.79.

The total remaining recoverable petroleum estimated to occur in the Midcontinent region is shown in Table 7.80 as about 7.1 billion barrels of oil, 3.0 billion barrels of natural gas liquids, and 90.0 trillion cubic feet of natural gas. More energy remains in natural gas than in oil. As the reserves/production ratio for oil is relatively low, production will continue to depend primarily upon reserve additions and will generally trend downward if reserve additions remain below production levels. Exploration will have to be concentrated on smaller fields than in the past. There is a substantial estimated undiscovered gas resource, but it is thought to occur mostly in the deep Anadarko basin, where exploration and development is very expensive.

Eastern Interior

Petroleum Geology

The Eastern Interior region includes Illinois, Indiana, western Kentucky, and Tennessee. In this region there are two main structural features that are petroleum productive, the Illinois basis (located in central and southern Illinois, southwestern Indiana, and western Kentucky) and the Cincinnati arch (located in eastern Indiana, central Kentucky, and western Tennessee). The Illinois basin is primarily an oil province, with relatively little gas produced, while the Cincinnati arch has been productive in both gas and oil.

The Illinois basin has produced oil for more than 100 years. The first major oil boom began in 1904 with the discovery of the Casey pool in southeastern Illinois. The basin is a Paleozoic, interior basin in which up to 13,000 to 14,000 feet of sediments range in age from Cambrian to Pennsylvanian. In the extreme southern part of Illinois and in Kentucky, Cretaceous and Tertiary rocks occur due to the encroachment of the Upper Mississippian embayment, but these sediments have poor petroleum potential.[64]

The early discoveries in the Illinois basin were followed by the development of shallow production along the La Salle anticlinal belt on the western shelf of the basin. In 1937, a second drilling boom started in the deep part of the basin, which resulted in the discovery, in 1938, of three national-class giant fields (Louden, Salem, and Clay City) and a fourth national-class giant (New Harmony) in 1939. These fields produce oil primarily from Mississippian sandstones, but Devonian limestones also are sometimes productive. The traps are local anticlines along the larger Du Quoin or La Salle fold structures. Local productivity is largely controlled by permeability variations, and thus some accumulations are partially stratigraphic. These older national-class giant fields are now in advanced stages of decline.

Some associated gas is present in Illinois basin fields as solution gas or (rarely) as a gas cap, but little gas is marketed. Other fields in the basin have been found in reefs and in local structural traps developed in Devonian and Silurian carbonates.

The Cincinnati arch is a structurally positive element extending northward across Tennessee and Kentucky between the Illinois basin on the west and the Appalachian basin on the east. On the north, the arch is split and its limbs constitute the southern margin of the Michigan basin. Oil and gas reservoirs lie at shallow depths along the arch, which contains a sequence of Paleozoic sediments ranging in thickness from less than 3,000 feet to more than 7,500 feet. Petroleum has been obtained from the province since 1829, when oil was gathered in Cumberland County, Kentucky. The national-class giant Trenton field, at the north end of the arch, was discovered in Ordovician dolomites in 1884. The peak oil production from the entire arch region was achieved at the turn of the century because of production from this field, which is now virtually exhausted. Some shallow gas was also produced from the Trenton field.

The oil fields of western Kentucky account for most of the state's oil output, while most of the gas is produced from the eastern portion of the state. (Kentucky gas production will be included in the Appalachians section.) The principal structural feature of the western Kentucky area is the 100-mile-long Rough Creek fault system. A number of small fields are associated with this structural trend and produce from Silurian and Devonian limestones and Mississippian and Pennsylvanian sandstones. The source beds are believed to be nearby shales.

During 1980, the number of exploration wells drilled in Illinois increased by 92 percent over the previous year, and the number of development wells increased by 71 percent. The increased drilling does not appear to have resulted from any new exploration trends, but rather from the extension of production in old areas and from in-fill drilling.[65] Wells have been deepened in established producing areas to test deeper Mississippian and Devonian formations, and in some cases excellent production has been established. The new production, which has been almost entirely oil with very little associated gas, has been concentrated in southeastern Illinois.

In Indiana, drilling increased by about two-thirds over the previous year. The drilling activity was centered in southwestern Indiana, where deeper drilling to Mississippian limestones and sandstones resulted mainly in oil discoveries with very little associated gas. The number of wells completed in Kentucky in 1980 increased by 42 percent over 1979. As the ratio of dry holes to total holes drilled remained about the same as in 1979, there was also a significant increase in the number of dry holes. Drilling also increased significantly in Tennessee in 1980, with exploration primarily for petroleum accumulations in Mississippian carbonates.[66]

Production, Reserves, and Resources

A historical sequence of the past decade in oil production in the Eastern Interior is shown in Tables 7.81 through 7.84. Oil production in Illinois dropped by 67 percent over the past decade to 17 million barrels in 1980. Reserves declined by 58 percent (to 113 million barrels), as 159 million more barrels of oil were produced than were added to proved reserves (see Table 7.81). Thus, both production and reserves have fallen significantly. With relatively low reserves/production ratios, future production will continue to depend in large part on future reserve additions.

Indiana oil production fell by 50 percent during the last decade and reserves declined by 44 percent, as 18 million more barrels of oil were produced than were added to reserves. In 1980, 4 million barrels were produced from a proved reserve estimated at 23 million barrels (see Table 7.82). In this mature area with quite low reserves/production ratios, production will continue to depend mostly upon reserve additions, and declining reserve additions will be closely followed by declining production.

Kentucky oil production fell by 46 percent over the past decade to 7 million barrels in 1980. Reserves declined by 52 percent, as 38 million more barrels were produced than were added to reserves. Reserves in 1980 were estimated to be 35 million barrels (see Table 7.83). Although both production and reserves experienced a significant decline in the 1970s, reserves increased in 1980. Such reserve increases will be necessary to sustain production in this mature area in which reserves/production ratios are quite low.

Total Eastern Interior region oil production fell by 61 percent during the past decade to 28 million barrels in 1980. Reserves declined by 56 percent, as 215 million more barrels of oil were produced than were added to reserves. Eastern Interior proved oil reserves in 1980 were estimated to be 171 million barrels (see Table 7.84). Reserves/production ratios have remained low; thus, future production will depend mostly upon reserve additions as in the 1970s. If reserve additions remain below production, production will continue to fall.

Inferred reserves for the region are estimated to be 86 million barrels. Table 7.85 contains estimates of undiscovered recoverable conventional oil resources in the region. The expected quantity of undiscovered oil in the region is 800 million barrels. The low and high estimates are not additive, but have been modified from U.S. Geological Survey calculations as 300 and 1,500 million barrels, respectively. It is evident from the table that the Illinois basin is considered more prospective for oil than the Cincinnati arch.

Table 7.86 indicates that the estimated ultimate oil recovery for the region is about 5.6 billion barrels. About 86 percent of the total recoverable oil estimated to have existed in the region is thought to already have been discovered. Given a relatively small undiscovered oil resource, it may be difficult to stabilize production in the long term.

The Eastern Interior region produces very little gas. Gas reserves are also quite small (about 0.010 trillion cubic feet, with inferred reserves of about 0.1 trillion cubic feet). Undiscovered recoverable gas resources have been projected as: Illinois basin, 0.1 to 2.1 trillion cubic feet with a mean (expected volume) of 0.6 trillion cubic feet; and Cincinnati arch, 0.1 to 1.9 trillion cubic feet with a mean (expected volume) of 0.7 trillion cubic feet.

Petroleum prospects remain favorable for the Eastern Interior region, but production is expected to continue its decline. The future trend will be toward in-fill drilling, deeper drilling, and a search for stratigraphic traps.

Michigan Basin

Petroleum Geology

The Michigan basin is an interior basin covering about 122,000 square miles of the southern peninsula of Michigan (see Figure 7.2). It contains about 15,000 feet of mostly Paleozoic sedimentary rocks. The sediments filling the basin are of Cambrian through Pennsylvanian age, with remnant Jurassic sediments also occurring. The Michigan basin has been a petroleum-producing province for over fifty years. Early production was primarily from Devonian and Mississippian formations, but more recently Ordovician and especially Silurian rocks have become the most important producers.

The Michigan basin subsided from Late Cambrian through Mississippian time. Early Silurian deposition was primarily of shales and carbonates; Middle and Late Silurian times were characterized by a greatly accelerated rate of subsidence and the formation of a thick evaporite sequence in the interior of the basin, with a belt of Niagaran pinnacle reefs around the edge. These reefs are thought to have grown in a broad open, shallow sea some distance from land on shoals surrounding the rapidly subsiding basin.[67]

Recent exploration in the Michigan basin has successfully concentrated on these Middle Silurian Niagaran pinnacle reefs. About 550 productive reefs have been found in the northern reef trend. Niagaran reefs currently supply 81 percent of Michigan's total oil production and 89 percent of its total gas production. These totals include the south-central reef area, which is not as productive as the northern reef trend but is shallower and therefore less expensive to drill. Exploration for Niagaran reefs is projected to continue at about the same level, with primary concentration on the northern reef belt.

There is increased exploration of the central basin area, where recent deeper drilling has encountered oil in Devonian rocks. Recent gas discoveries in Ordovician sandstones below 10,000 feet also have increased exploration interest in the deeper parts of the basin.[68] Higher petroleum prices are expected to promote continued drilling of these Devonian and Ordovician prospects in the deeper central basin area. General drilling activity is expected to increase.[69]

The fields discovered in the Michigan basin are relatively small, as is normal for an interior basin. The basin contains a somewhat higher proportion of gas than occurs in some other interior basins, however.

Production, Reserves, and Resources

The region's oil-producing history during the past decade is shown in Table 7.87. Both production and reserves have increased significantly. In 1980, 37 million barrels of oil were produced from a proved reserve estimated at 205 million barrels. Reserves are almost four times larger than at the beginning of the decade, as 153 million barrels more oil were added to reserves than were produced. The increase in reserves has permitted an increase in production, which expanded by a factor of three. The low reserves/production ratios at the beginning of the last decade made increased reserve additions necessary for increased production.

Inferred oil reserves for the region are estimated to be 820 million barrels. Undiscovered recoverable conventional oil resources are estimated to range from 300 to 2,700 million barrels with an expected recovery (mean) of 1,100 million barrels.

Table 7.88 indicates that the estimated ultimate oil recovery for the Michigan basin is about 3.0 billion barrels. About 63 percent of the total oil estimated to have existed in the basin has already been discovered. A reasonable amount remains to be found but is expected to occur in relatively small fields. There is also a substantial inferred reserve, which will aid production as it is developed.

The region's natural gas production history is shown in Table 7.89. Reserves were increased by about 8.5 times during the past decade, as 1.174 trillion cubic feet more gas was added to reserves than was produced. Reserves/production ratios were low in the late 1960s, so increased production depended on increasing reserves. New gas fields were discovered and production increased by a factor of 5. In 1980, 0.160 trillion cubic feet of gas was produced from a reserve of 1.329 trillion cubic feet. With a reserves/production ratio of 8:1, gas production stabilization in the near term looks favorable. Longer-term production will depend upon the discovery of additional reserves.

Inferred gas reserves for the Michigan basin are estimated to be 1.3 trillion cubic feet. Undiscovered recoverable conventional gas resources are estimated to range from 1.8 to 10.9 trillion cubic feet, with an expected recovery (mean) of 5.1 trillion cubic feet.

The estimated ultimate gas recovery from the Michigan basin is about 9.1 trillion cubic feet, as shown in Table 7.90. It can be seen from the table that 56 percent of Michigan basin gas is estimated to be as yet undiscovered. Thus, the basin is considered to have favorable natural gas prospects, but the new fields will probably be relatively small and/or deep and therefore expensive to find and develop.

The Energy Information Administration estimated 1980 natural gas liquids production in Michigan at 14 million barrels from a reserve of 112 million barrels. Thus, in 1980, there were 87.5 barrels of natural gas liquids recovered per million cubic feet of produced natural gas. On that basis, the ultimate recovery of natural gas liquids from the region is calculated as about 0.4 billion barrels, as shown in Table 7.91.

The total remaining recoverable petroleum estimated to occur in the Michigan basin is shown in Table 7.92 as about 2.1 billion barrels of oil, 0.3 billion barrels of natural gas liquids, and 7.7 trillion cubic feet of natural gas. The basin does not contain a large amount of petroleum, but it is still prospective. Future exploration will be more expensive as deeper and smaller fields become exploration objectives, which will be made possible by rising petroleum prices.

The Appalachians

Petroleum Geology

The Appalachian region includes New York, Pennsylvania, West Virginia, Virginia, Ohio, and eastern Kentucky. The Appalachian basin is a large, Paleozoic, intracontinental composite basin containing great thicknesses of sediment (see Figure 7.2). Petroleum accumulations occur in anticlinal folds and in local stratigraphic traps caused by variations in reservoir-rock porosity. The major oil and gas deposits are found in sediments of Cambrian through Pennsylvanian age.

The oil industry in the United States began in the Appalachian basin, at Titusville, Pennsylvania, in August 1859, when the Drake well demonstrated that oil could be obtained by drilling through rock. By chance, this discovery well was drilled in an area where the reservoir sands were very shallow and the oil light and sulfur free. This encouraged rapid and large-scale production and provided the impetus for the expansion of the industry to other fields and other states.

In Pennsylvania, production is derived from Silurian-through Pennsylvanian-age reservoirs. The Upper Devonian sandstones are productive in the world-class giant Bradford oil field, which extends into New York. Discovered in 1871, the field consists of two broad anticlines with gentle dips that plunge southward. It covers an area of about 85,000 acres and has been waterflooded. The source rocks are thought to be local carbonaceous shales. Below the Upper Devonian reservoirs, Lower Devonian Oriskany gas sandstones and Silurian Tuscarora and Medina gas and oil sandstones also are productive. Important deeper horizon fields include: Summit, Leidy, and Ellisburg. The rocks beneath the Silurian have not been evaluated in Pennsylvania, but these deeper formations are considered to have petroleum potential.

Exploratory drilling increased by 91 percent in Pennsylvania in 1980, and development drilling climbed by 6 percent. Almost 90 percent of 1980 gas production was from shallow Late Devonian, Mississippian, and Pennsylvanian reservoirs, and deeper drilling for Middle Devonian and older formations increased by 102 percent. The deepest producing well in the state is 11,458 feet in the Lower Silurian Tuscarora sandstone. The deepest test well drilled in the Appalachian basin went to 21,450 feet in Upper Cambrian rocks.[70]

In the southwestern part of New York several small oil fields with Devonian-age sandstone reservoirs have been found in anticlinal structures. New York oil and gas production is primarily from Devonian and Silurian rocks, with oil mostly from the Upper Devonian in areas adjacent to Pennsylvania. Gas is primarily derived from the Lower Devonian Oriskany and Lower Silurian Medina sandstones with some Ordovician Trenton limestone production.[71] Devonian reefs are also productive of gas in New York.

The petroleum-producing area of West Virginia is a continuation of the Pennsylvania oil region. The main structural feature is the Warfield–Chestnut Ridge regional anticline, which extends from northern Pennsylvania through West Virginia and into eastern Kentucky. A number of secondary structural features are associated with this anticline. The principal reservoirs (primarily sandstones) are of Early Mississippian, Late Devonian, Early Devonian, and Late Silurian age. Recent exploratory efforts have been concentrated in shallow Upper Devonian sandstones. There has been deeper drilling (primarily for gas) in West Virginia, but deep drilling decreased slightly in 1980. Only about 5 percent of the wells drilled in the state in 1980 tested the deeper formations.

The Appalachian basin extends into Ohio where oil and gas are found in various Mississippian and Permian sandstones on the flanks of gentle folds. There is also deeper petroleum production from the Lower Devonian Oriskany, Lower Silurian Medina or Clinton sandstones, and Cambrian dolomites. Some of the traps in Ohio result from ancient shoreline deposition along an unconformity that included channels and deltas and are thus stratigraphic in nature. The Cambrian–Ordovician sediments in the deeper part of the basin in eastern Ohio are relatively unexplored and may have petroleum potential. Drilling decreased in Ohio in 1980 by 3.7 percent, but 1980 was still the second most active drilling year in the past decade. There were almost four times as many gas wells completed as oil wells, but the majority of the successful wells produced both oil and associated gas. Most of the wells were drilled to the Silurian Clinton-Medina sandstone; wells drilled to the Mississippian Berea sandstone were the second most numerous.[72]

The Chestnut Ridge regional anticline continues into eastern Kentucky, where the petroleum accumulations are related to this structural feature. The major producing reservoirs are of Early Devonian and Silurian age. As almost all of Kentucky's gas production is derived from the eastern part of the state, which is geologically related to the Appalachian region, Kentucky gas production and reserve data are included here.

Drilling in Virginia in 1980 increased slightly over the previous year but continued at a relatively low level. Formations tested included the Mississippian Berea sandstone, the Devonian Oriskany sandstone, the Silurian Clinch sandstone, and the Ordovician Trenton limestone. Almost all of Virginia's petroleum production comes from the counties that form the southwestern tip of the state. In this area, the geology is similar to that of West Virginia, eastern Kentucky, and other Appalachian areas.

In general, the Appalachian basin is in the early stages of an important new exploratory trend. Natural gas potential may exist in the eastern portion of the eastern overthrust belt, where a buried wedge of 10,000 to 20,000 feet of clastic sedimentary rocks extending from New York to Alabama may be present. If these rocks are not metamorphosed, an untested portion of the basin exists with a very large volume of prospective rocks. Recent seismic surveys have indicated that past thrust faulting has moved the crystalline rocks of the Blue Ridge and Piedmont Plateau westward for at least 100 miles, burying a large sequence of the sedimentary rocks of the eastern overthrust belt. The overlying crystalline rock cover ranges from about 5,000 feet thick along its western edge to more than 10,000 feet thick in the east. The burial occurred near the end of the Paleozoic era when the Blue Ridge and Piedmont Plateau were thrust into their present positions along the eastern slope of the Appalachian Mountains. The overthrust area from central Virginia southward is believed to have deep gas potential, but the overthrust region to the north may have been subjected to too much heat during the maturation period to contain even thermal gas.[73]

Deep gas prospecting has been confined to the exposed thrust region in front (to the west) of the concealed overthrust belt, where the drilling depths are not as great and it is not necessary to penetrate thousands of feet of crystalline rocks before reaching the sediments. A number of gas fields have been found in the exposed thrust region in the Silurian Tuscarora and Lower Devonian Oriskany sandstones, and it is probable that the concealed portion of the overthrust belt will eventually be tested also.[74]

Production, Reserves, and Resources

A historical sequence of the region's oil situation is given in Tables 7.93 through 7.96. As shown in Table 7.93, Pennsylvania oil production dropped by 50 percent in the 1970s to 2 million barrels. Reserves remained about the same at 54 million barrels. Reserves/production ratios are relatively high, indicating that the modest production can be sustained in the near term.

In West Virginia, 23 million more barrels of oil were produced than were added to reserves in the last decade, causing reserves to decline by 43 percent to 30 million barrels. Production also fell, to 2 million barrels (see Table 7.94). Reserves/production ratios indicate that additional reserve drawdown is possible to stabilize the modest production in the near term.

Ohio oil production, reserves, and reserves/production ratios have remained relatively stable for the period. There was a recorded drop in reserves in 1980, but there are indications that this may be mostly statistical. Reserves/production ratios indicate that production may be sustained in the short term, after which additional reserve additions will be needed. Production in 1980 was 12 million barrels from a reserve reported as 116 million barrels (see Table 7.95).

In the Appalachian region 35 million more barrels of oil were produced than were added to reserves during the past decade, and reserves declined by 15 percent to 200 million barrels (see Table 7.96). Production fell by 11 percent to 16 million barrels. Reserves/production ratios indicate that, in the near term, it will be possible to support production by drawing down reserves. In the longer term Appalachian region oil production will depend upon reserve additions.

Inferred oil reserves for the region are estimated to be 100 million barrels. The undiscovered recoverable conventional oil resources for the Appalachian basin are estimated to range from 100 to 1,500 million barrels with an expected recovery (mean) of 600 million barrels.

Table 7.97 indicates that the projected ultimate oil recovery for the region is about 3.7 billion barrels. Of the total oil estimated to occur, 84 percent has already been discovered. The remaining oil will be in smaller fields and in more subtle traps and will be more difficult and expensive to find.

The past decade of natural gas production in the region is shown in Tables 7.98 through 7.103. As 0.210 trillion cubic feet more gas was added to reserves in the past decade than was produced, reserves generally increased over the decade in Pennsylvania to 1.016 trillion cubic feet (see Table 7.98). Gas production also generally increased, but then fell to 0.068 trillion cubic feet in 1980. Reserves/production ratios are relatively high, and thus near-term production could be supported by drawing on reserves if necessary.

West Virginia gas reserves increased to 2.422 trillion cubic feet during the past decade, as 0.313 trillion cubic feet more gas was added to reserves than was produced. Production, however, declined by 29 percent to 0.151 trillion cubic feet (see Table 7.99). Thus, reserves/production ratios increased to 16:1, a relatively high ratio that indicates a reserve available if an increase in production is required.

Ohio gas reserves increased by a factor of almost 4, to 1.508 trillion cubic feet during the decade, as 1.116 trillion cubic feet more gas was added to reserves than was produced. Production increased by over 2.5 times to 0.137 trillion cubic feet (see Table 7.100). Reserves/ production ratios also increased (to 11:1).

New York gas reserves increased during the decade by almost six times to 0.161 trillion cubic feet, as 0.133 trillion cubic feet more gas was added to reserves than was produced. Production almost tripled (to 0.014 trillion cubic feet) but is still quite low. Reserves/production ratios increased to 12:1 (see Table 7.101).

Kentucky gas reserves declined by 30 percent during the 1970s (to 0.605 trillion cubic feet), as 0.265 trillion cubic feet more gas was produced than was added to reserves. As Table 7.102 shows, production dropped to 0.060 trillion cubic feet, which represents a 24 percent decline during the decade. Reserves/production ratios fell to 10:1, indicating that production will depend more upon reserve additions in the future than in the recent past.

Appalachian gas reserves as a whole increased to 5.712 trillion cubic feet during the decade, as 1.512 trillion cubic feet more gas was added to reserves than was produced. Production increased by about 3 percent (to 0.430 trillion cubic feet), and reserves/production ratios increased to 13:1 (see Table 7.103).

Inferred reserves for the region are estimated to be 5.4 trillion cubic feet. Table 7.104 contains estimates of undiscovered recoverable gas resources in the Appalachian region. The expected volume of undiscovered gas is 20.2 trillion cubic feet. The low and high estimates are not additive, but have been calculated as 6.4 and 45.8 trillion cubic feet by the U.S. Geological Survey. The Potential Gas Committee estimated that the eastern states may contain 11 trillion cubic feet of undiscovered gas below 15,000 feet.

The estimated ultimate gas recovery from the region is about 65.9 trillion cubic feet, as shown in Table 7.105. About 69 percent of the estimated total Appalachian region gas has already been discovered. More gas has been produced in the region than is estimated to remain undiscovered.

Appalachian natural gas liquids production for 1980 was estimated by the Energy Information Administration as 9 million barrels from a reserve of 122 million barrels. Thus, in 1980, there were 21 barrels of natural gas liquids produced per million cubic feet of produced gas. On that basis, the ultimate recovery of natural gas liquids in the region is calculated as about 1.2 billion barrels, as shown in Table 7.106.

The total remaining recoverable petroleum estimated to occur in the Appalachian region is shown in Table 7.107 as about 0.9 billion barrels of oil, 0.4 billion barrels of natural gas liquids, and 31.3 trillion cubic feet of natural gas. On a Btu basis there is more energy estimated to remain in the region in gas than in oil. In the near term it will be possible to support gas production with reserve drawdown, and there is a sizable inferred gas reserve that can be developed and a significant undiscovered gas potential. However, these gas prospects will be deeper and thus more expensive than those developed in the past.

Atlantic Coast

The Atlantic Coast province includes the Atlantic coastal plain and the Atlantic continental shelf and slope, extending from New England to eastern Florida (see Figure 7.2). There has been comparatively little petroleum production in this region to date. There is currently relatively minor oil production from Lower Cretaceous reefs onshore in southern Florida.[75] In spite of this lack of significant petroleum production, the large Atlantic Coast region is thought to have considerable petroleum potential, especially offshore. A seaward-thickening wedge of sedimentary rocks of Mesozoic and Tertiary age underlies the Atlantic coastal plain and extends outward beneath the continental shelf and slope. Most of the sediments were deposited in transitional or shallow-marine environments that offered a mix of potential source and reservoir formations and provided good potential for stratigraphic traps.[76]

The offshore eastern continental margin can be divided into a northern physiographic segment, extending from New England to Cape Hatteras, North Carolina, and a southern segment between Cape Hatteras and the Florida Keys. The northern segment is characterized by a relatively smooth, gently dipping shelf extending seaward to a water depth of 300 to 650 feet. A break at this depth marks the beginning of the more steeply inclined continental slope. The southern segment of the margin differs from the northern in that the Blake Plateau intervenes between the continental shelf and slope.[77]

A series of pull-apart basins is located along the Atlantic continental margin, occupying the intermediate crustal zone between the thick continental crust and the thin oceanic crust. The sediments are predominantly clastics, but carbonate banks also occur. To date, explored pull-apart basins in other regions of the world have displayed low petroleum productivity, but these offshore eastern basins are important to the petroleum future of the United States because they have been estimated to contain 20 percent of undiscovered domestic offshore oil and 14 percent of undiscovered domestic offshore gas.

Georges Bank is the most southwestern of a series of banks on the North Atlantic continental shelf that parallel the coast between Newfoundland and Nantucket Island. The Georges Bank basin is a structural trough about 200 miles long by 80 miles wide that contains in excess of 26,000 feet of Mesozoic and Cenozoic sediments. Wells drilled in the basin have penetrated rocks with low organic content that appear to reflect deposition under oxidizing conditions in an open shelf environment. These conditions are not conducive to the preservation of organic matter and result in poor source rocks. Sediments deposited in the more reducing environment of topographic lows within an open shelf basin or a slope environment may contain larger amounts of preserved organic material. Basinal shales have been identified in the vicinity of Oceanographer Canyon on Georges Bank, which may be a potentially oil-prone area.[78] Possible petroleum traps in the Georges Bank basin include sediments draped over large basement blocks and carbonate banks. Stratigraphic traps are also possible in areas where clastic rocks grade to carbonates.

A sale of federal leases in the Georges Bank basin was conducted in December 1979, but drilling was delayed until mid-1981 because of court challenges. One well, which was dry, has been completed, and three more wells are in the process of being drilled. Because of its size, accessibility, and moderate water depths (less than 260 feet), Georges

Bank may play an important role in the nation's petroleum future. If prospects are to be favorable, however, areas must be found that are adjacent to mature source rocks.

The structure of the Mid-Atlantic margin represents a southward continuation of the trends evident beneath Georges Bank. The Mesozoic and Cenozoic sedimentary sequence thickens considerably off the coast of southern New Jersey, in the Baltimore Canyon trough. On the coastal plain of New Jersey, Delaware, Maryland, and Virginia, more than 6,500 feet of Cenozoic and Mesozoic sediments have been penetrated by many wells. These sedimentary rocks unconformably overlie a basement of Paleozoic and older metamorphic rocks. Sedimentary rocks in the Baltimore Canyon trough thicken seaward to about the base of the continental shelf, where they begin to thin. Near the seaward margin of the shelf, a carbonate platform, which formed throughout much of Jurassic time, occurs over basement rocks. The trough appears to contain in excess of 40,000 feet of sedimentary rocks.

Data from test wells indicate that Late Cretaceous and Tertiary rocks contain organic material but are thermally immature in respect to petroleum generation. The younger sedimentary rocks on the outer portions of the continental shelf and upper part of the continental slope also have a relatively low potential for the generation of liquid petroleum, but significant amounts of natural gas could have been generated by older rocks at depths below about 9,800 feet. However, there is also a deterioration of sandstone porosity below 9,800 feet. This, along with a possible lack of lateral continuity of Jurassic sandstone beds, will make exploration difficult and suggests a low probability of the discovery of giant gas fields on the Mid-Atlantic shelf. The conditions for the entrapment of natural gas in Jurassic sandstones are favorable, but the fields may be small because of thin reservoir beds and facies changes in a seaward direction. There are thicker and more porous rock units in the overlying Cretaceous sequence into which vertical migration of gas may have occurred through faults that cut Jurassic and Early Cretaceous strata along the shelf margin.[79]

Also of interest for petroleum exploration, but in deep water, is the Jurassic and Early Cretaceous carbonate bank buildup beneath the outer continental shelf and upper continental slope. The Jurassic paleo-shelf margin was dominated by carbonate rocks that may contain potential petroleum traps in fore-reef, reef, and back-reef facies. It is possible that the basinal shale facies, which interfingers with the carbonate rocks, is a potential source for liquid petroleum. The back-reef environment has poor source-rock potential, but the slope sediments are likely to have a more marine character and thus preserve larger amounts of organic material. Adequate reservoirs will depend upon the amount of secondary porosity that has formed within the reef platforms by subaerial exposure and weathering. The nearest test well indicates that the organic debris appears terrigenous and is likely to produce gas.[80]

Of the 26 exploratory wells drilled in the Baltimore Canyon trough, so far, 21 have been dry and 5 have yielded significant recoveries of natural gas. A commercial-size field has yet to be established.

South of the Baltimore Canyon trough, the basement rock rises gradually to the southwest and then descends again to form the Southeast Georgia Embayment. The embayment comprises about 13,000 feet of Paleozoic through Cenozoic sediments. Following two federal lease sales, 6 dry exploratory wells have been drilled in the embayment, and indications are that mature organic-rich source rocks are absent. This, added to the fact that much of the embayment area is under relatively deep water (in excess of 650 feet), will tend to inhibit further drilling.

The Blake Plateau trough is located about 140 miles off the coast of Georgia and Florida. The sedimentary rocks in this area may be more than 20,000 feet in thickness but are beneath 1,500 to 6,000 feet of water. Because of the deep water overlying the trough, it may be some time before this area can be considered a serious candidate for development.

As Table 7.108 shows, the expected volume of undiscovered oil in the Atlantic Coast region is 5,700 million barrels. The Mid-Atlantic (Baltimore Canyon) area is estimated to have the best oil prospects in the region. The low and high undiscovered oil resource estimates are not additive and range from 1,100 million barrels to 12,900 million barrels offshore and from 100 million barrels to 800 million barrels onshore.

The expected volume of undiscovered gas in the region is 23.6 trillion cubic feet, as shown in Table 7.109. As in the case of oil, the Mid Atlantic area is considered to have the best natural gas prospects. The nonadditive low and high undiscovered gas estimates for the region are 9.2 and 42.8 trillion cubic feet offshore and 0 to 0.4 trillion cubic feet onshore. The expected volume of undiscovered recoverable natural gas liquids has been estimated to be negligible onshore and 800 million barrels offshore.

Explored pull-apart basins have exhibited low petroleum content, and exploration so far in the offshore Atlantic coastal basins has not produced a commercial field. Although petroleum prospects remain favorable in the Atlantic Coast region, the better locations appear to be further from shore and in relatively deep waters. If fields exist in such environments, they will have to be large to warrant commercial development and the petroleum derived will be expensive to recover. Each individual assessment area has 0 for a low estimate of recovery because economically recoverable petroleum has yet to be proved, but the overall likelihood of discovery for all the areas combined is greater than 0 (as reflected in the low assessments for the region as a whole). It must be remembered that petroleum production is still not assured from any of these basins.

Summary

Domestic oil and gas production peaked during the past decade. A historical sequence of petroleum production during this period, shown in Table 7.110, is important to the understanding of the exploration and production process and of the essential nature of reserves. United States oil production decreased from a high in 1970 of 3,328 million barrels to 2,975 million barrels in 1980 (about 11 percent). During this same period proved reserves declined by 31 percent to 27,046 million barrels. Since 1970, reserves have been drawn down to support production. In 1970 the reserves in the super-giant Prudhoe Bay field in Alaska were added to domestic proved reserves. If Prudhoe Bay had not been discovered and developed, domestic production would have fallen by 27 percent and reserves by 56 percent by 1980; the reserves/production ratio would be 7:1 (compared to the actual 9:1). The importance of super-giant fields even to as large a producing nation as the United States cannot be overemphasized. U.S. oil production will continue to depend more upon reserve additions in the near term as reserves/production ratios decline.

Inferred oil reserves (projected growth of known fields) for the United States are estimated to be 26,300 million barrels. Undiscovered recoverable conventional oil resources are estimated to range from 64,300 to 105,100 million barrels with an expected recovery (mean) of 82,600 million barrels. Table 7.111 indicates that the estimated ultimate oil recovery for the United States is about 259.6 billion barrels. According to the table, 68 percent of the total oil estimated to have existed in the United States has already been discovered, and more has been produced than is projected to remain undiscovered. It is obvious, however, that a considerable domestic oil resource remains to be utilized.

The nation's natural gas situation for the past decade is shown in Table 7.112. Domestic gas production peaked in 1973 at 22.605 trillion cubic feet and has since declined by 14 percent to 19.333 trillion cubic feet. Gas reserves have continuously declined during the 1970s to 188.161 trillion cubic feet (by 31 percent). Reserves/production ratios have fallen, indicating that reserve drawdown cannot support production in the near term to the extent that was achieved in the 1970s. Future gas production will, therefore, depend more upon reserve additions.

Inferred gas reserves for the United States are estimated to be 174.4 trillion cubic feet. Undiscovered recoverable conventional gas resources are estimated to range from 474.6 to 739.3 trillion cubic feet, with an expected recovery (mean) of 593.8 trillion cubic feet.

The estimated ultimate gas recovery from the United States is about 1,553.7 trillion cubic feet, as shown in Table 7.113. Although 62 percent of the total natural gas estimated to have existed in the United States has already been discovered, a considerable undiscovered gas resource remains.

The Energy Information Administration estimated 1980 natural gas liquids production in the United States at 731 million barrels from a reserve of 6,728 million barrels. Thus, the average production of gas liquids in the United States in 1980 was 37.8 barrels per million cubic feet of produced natural gas. On that basis, the ultimate recovery of natural gas liquids in the United States is calculated at about 53.6 billion barrels (see Table 7.114).

The total remaining recoverable petroleum in the United States is estimated as about 135.9 billion barrels of oil, 31.0 billion barrels of natural gas liquids, and 956.4 trillion cubic feet of natural gas, as shown in Table 7.115. There is a larger remaining energy resource in natural gas than in oil.

The oil and gas activity in the United States in 1980 is shown in Tables 7.116 and 7.117.[81] As the tables indicate, the best oil prospects are thought to be in the Rocky Mountains and Great Plains province. An estimated 50 to 70 percent of this undiscovered oil is on federal lands, however. Alaska also has considerable oil potential—most of it also on federal lands. The greatest gas potential is in the Gulf Coast and Rocky Mountains. Some areas, such as the Eastern Interior province, are being heavily explored for the relatively small amount of petroleum thought to remain. Obviously further drilling in lightly explored regions with substantial prospects is more likely to result in significant additions to domestic reserves than heavy drilling in very mature areas with relatively small volumes of remaining petroleum. This is not to suggest that drilling in the mature areas should be stopped, for the prospects there are economically viable (at the higher petroleum prices). Furthermore, any petroleum found, no matter how small the field, is of value if produced. On the other hand, large numbers of wells do not always indicate that significant exploration is taking place; the location of the wells is of primary importance.

In the United States, the unique phenomenon of private ownership of land and mineral resources has allowed a very thorough exploration for petroleum. Offshore petroleum, however, (an estimated 35 percent of undiscovered oil and 28 percent of undiscovered gas) is 95 percent under federal ownership. Onshore in the lower forty-eight states, 30 percent of the undiscovered oil and 25 percent of the undiscovered gas are estimated to be on federal lands.[82] Thus, the federal government will play a very important role in future domestic petroleum production.

During 1980, 1,340 new field discoveries were reported, an increase of 15.3 percent over 1979. These new discoveries are estimated to contain ultimate reserves of 504 million barrels of oil and condensate and 4.1 trillion cubic feet of gas. This is considerably less than domestic oil and gas production in 1980. The small amount of petroleum discovered for such a large number of new fields reflects the completion of more marginal wells. In 1980 one in 5.2 new field exploratory holes was completed as a producer well, but only 1 in 180 exploratory holes (less than 1 percent) discovered a field estimated to contain ultimate reserves in excess of 1 million barrels of oil or 0.006 trillion cubic feet of gas. In 1980 there were no domestic fields found that were estimated to contain in excess of 50 million barrels of oil equivalent and only 1 field found was in the 25 to 50 million barrels of oil equivalent range.[83] If the general oil and gas production declines experienced in the 1970s are to be reversed, larger fields will have to be found, and to find larger fields frontier areas will have to be drilled. Such drilling will not assure increased production, but the lack of drilling will assure continued production declines.

TABLE 7.1   Alaska Oil Production History (in million barrels)

Year	Production	Reserves	Reserves/Production Ratio
1969	74	432	6 : 1
1970	83	10,149	122 : 1
1971	78	10,116	130 : 1
1972	73	10,096	138 : 1
1973	72	10,112	140 : 1
1974	71	10,094	142 : 1
1975	70	10,037	143 : 1
1976	63	9,786	155 : 1
1977	169	9,616	57 : 1
1978	448	9,247	21 : 1
1979	511	8,924	17 : 1
1980	591	8,799	15 : 1
Total	2,303		

Reserve additions for the decade totaled 10,596 million barrels.

Sources: American Petroleum Institute, American Gas Association, and Canadian Petroleum Association, *Reserves of Crude Oil, Natural Gas Liquids, and Natural Gas in the United States and Canada as of December 31, 1979,* vol. 34 (Washington, D.C., 1980); and Energy Information Administration, Office of Oil and Gas, *Principal Findings of the U.S. Crude Oil, Natural Gas, and Natural Gas Liquids Reserves,* 1980 Annual Report (Washington, D.C.: Department of Energy, 1981). The reserve estimates for 1980 were derived from American Petroleum Institute, American Gas Association, and Energy Information Administration figures.

TABLE 7.2 Estimated Undiscovered Recoverable Conventional Oil Resources of Alaska (in million barrels)

Region	Offshore (shelf 0–650 feet, slope below 650 feet)			Onshore	
Gulf of Alaska	0–2,500	(500)[a]	Shelf	0–800	(200)
	0–1,600	(300)	Slope		
Cook Inlet– Alaska peninsula	100–2,000	(400)	Shelf	100–1,800	(600)
Bristol Bay basin	0–1,200	(200)	Shelf	0–600	(100)
St. George basin	0–2,200	(400)	Shelf		
Navarin basin	0–3,700	(800)	Shelf		
	0–600	(100)	Slope		
Norton basin	0–900	(200)	Shelf		
Hope basin[b]	0–100	(0)	Shelf		
Central Chukchi Sea basin[b]	0–3,300	(600)	Shelf		
North Chukchi	0–4,200	(800)	Shelf		
Sea basin[b]	0–1,100	(200)	Slope		
Arctic Slope	1,900–16,700	(7,000)	Shelf	1,200–16,400	(6,000)
(Beaufort Sea[b])	0–3,600	(800)	Slope		
Mean total		(12,300)			(6,900)

[a] The low estimate corresponds to a 95 percent probability of more than that amount and the high estimate to a 5 percent probability of more than that amount. The statistical mean (in parentheses) is the expected quantity (both offshore and onshore) of the resource and is additive.

[b] These quantities are only recoverable if the technology is developed to permit their exploitation beneath the Arctic pack ice.

Source: Modified from G. L. Dolton et al., *Estimates of Undiscovered Recoverable Conventional Resources of Oil and Gas in the United States,* Circular 860 (Reston, Va.: U.S. Geological Survey, 1981), pp. 22–25 and 76–79.

TABLE 7.3 Estimated Ultimate Alaska Oil Recovery (in billion barrels)

Cumulative production (to the end of 1980)	2.463
Proved reserves	8.799
Inferred reserves (growth)	5.0
Undiscovered recoverable resources	19.2
Total Alaskan oil	35.462

TABLE 7.4   Alaska Natural Gas Production History (in trillion cubic feet)

Year	Production	Reserves	Reserves/Production Ratio
1969	.087	5.2	60:1
1970	.145	31.1	214:1
1971	.154	31.4	204:1
1972	.146	31.5	216:1
1973	.130	31.6	243:1
1974	.144	31.9	222:1
1975	.155	32.1	207:1
1976	.224	31.9	142:1
1977	.187	31.8	170:1
1978	.214	31.6	148:1
1979	.225	31.9	142:1
1980	.213	33.0	155:1
Total	2.024		

Reserve additions for the decade totaled 29.737 trillion cubic feet.

Source: See Table 7.1.

TABLE 7.5 Estimated Undiscovered Recoverable Conventional Natural Gas Resources of Alaska (in trillion cubic feet)

Region	Offshore (shelf 0–650 feet, slope below 650 feet)		Onshore
Gulf of Alaska	0–15.0 (3)[a]	Shelf	0–1.7 (0.3)
	0–6.6 (1.1)	Slope	
Cook Inlet	0.7–7.9 (2.6)	Shelf	1.1–9.3 (3.7)
Alaska peninsula	0–1.9 (0.3)	Slope	
Bristol Bay basin	0–2.6 (1)	Shelf	0–2.3 (0.5)
St. George basin	0–11.7 (2.4)	Shelf	
Navarin basin	0–22.1 (5.2)	Shelf	
	0–2.8 (0.4)	Slope	
Norton basin	0–5.5 (1.2)	Shelf	
Hope basin[b]	0–1.8 (0.3)	Shelf	
Central Chukchi Sea basin[b]	0–15.3 (3)	Shelf	
North Chukchi Sea basin[b]	0–17.7 (3.4)	Shelf	
	0–4.9 (1.1)	Slope	
Arctic Slope	9.4–85.3 (35)	Shelf	7–84.8 (31.8)
(Beaufort Sea[b])	0–17.8 (4.3)	Slope	
Yukon-Koyukuk			0–0.6 (0.1)
Mean total	(64.3)		(36.4)

[a] See Table 7.2 for an explanation of estimates.

[b] These quantities are recoverable only if the technology is developed to permit their exploitation beneath the Arctic pack ice.

Source: See Table 7.2.

TABLE 7.6   Estimated Ultimate Alaska Natural Gas Recovery (in trillion cubic feet)

Cumulative production (to the end of 1980)	2.267
Proved reserves	33.0
Inferred reserves (growth)	5.8
Undiscovered recoverable resources	100.7
Total Alaskan natural gas	141.767

TABLE 7.7   Estimated Remaining Alaska Petroleum

	Oil (billion barrels)	Gas Liquids (billion barrels)	Natural Gas (trillion cubic feet)
Proved reserves	8.799	.400	33.0
Inferred reserves (growth)	5.0	.070	5.8
Undiscovered resources	19.2	3.4	100.7
Total remaining petroleum	32.999	3.870	139.5

TABLE 7.8   California Oil Production History (in million barrels)

Year	Production	Reserves	Reserves/Production Ratio
1969	375	4,243	11:1
1970	372	3,984	11:1
1971	358	3,706	10:1
1972	347	3,554	10:1
1973	336	3,488	10:1
1974	323	3,557	11:1
1975	322	3,648	11:1
1976	326	3,590	11:1
1977	350	3,632	10:1
1978	347	3,471	10:1
1979	352	3,645	10:1
1980	360	3,787	11:1
Total	4,168		

Reserve additions for the decade totaled 3,337 million barrels.

Source: See Table 7.1.

TABLE 7.9   Estimated Undiscovered Recoverable Conventional Oil Resources of California (in million barrels)

Region	Offshore (shelf 0-650 feet, slope below 650 feet)			Onshore	
Inner basins	100-900	(400)[a]	Shelf		
	0-1,000	(200)	Slope		
Outer basins and ridges	0-100	(0)	Shelf		
	0-2,500	(500)	Slope		
Santa Barbara Channel	100-1,600	(600)	Shelf		
	100-2,100	(700)	Slope		
Santa Maria basin	0-800	(200)	Shelf	0-500	(200)
	0-2,200	(500)	Slope		
Santa Cruz basin	0-600	(100)	Shelf	0-400	(200)
	0-500	(100)	Slope		
Bodega basin	0-200	(0)	Shelf		
	0-200	(0)	Slope		
Point Arena basin	0-200	(0)	Shelf		
	0-400	(100)	Slope		
Eel River basin	0-200	(0)	Shelf		
	0-200	(0)	Slope		
San Joaquin basin				500-4,400	(1,800)
Los Angeles basin				200-1,500	(700)
Ventura basin				200-1,200	(500)
Central Coastal basins				0-400	(200)
Mean total		(3,400)			(3,600)

[a] For an explanation of estimates, see Table 7.2.
Source: See Table 7.2.

TABLE 7.10   Estimated Ultimate California Oil Recovery (in billion barrels)

Cumulative production (to the end of 1980)	18.811
Proved reserves	3.787
Inferred reserves (growth)	2.8
Undiscovered recoverable resources	7.0
Total California oil	32.398

TABLE 7.11   California Natural Gas Production History (in trillion cubic feet)

Year	Production	Reserves	Reserves/Production Ratio
1969	.665	6.684	10:1
1970	.634	6.099	10:1
1971	.589	5.517	9:1
1972	.514	5.045	10:1
1973	.478	4.920	10:1
1974	.357	4.862	14:1
1975	.332	5.155	16:1
1976	.313	4.964	16:1
1977	.332	4.801	14:1
1978	.305	4.721	15:1
1979	.331	4.648	14:1
1980	.299	5.010	17:1
Total	5.149		

Reserve additions for the decade totaled 2.810 trillion cubic feet.

Source: See Table 7.1.

TABLE 7.12   Estimated Undiscovered Recoverable Conventional Natural Gas Resources of California (in trillion cubic feet)

Region	Offshore (shelf 0–650 feet, slope below 650 feet)		Onshore
Inner basins	0.1–0.8 (0.3)[a]	Shelf	
	0–0.9 (0.2)	Slope	
Outer basins and ridges	0–0.1 (0)	Shelf	
	0–4.4 (0.9)	Slope	
Santa Barbara Channel	0.2–3.1 (1.0)	Shelf	
	0.1–4.4 (1.5)	Slope	
Santa Maria basin	0–0.7 (0.2)	Shelf	0–0.4 (0.2)
	0–2.0 (0.4)	Slope	
Santa Cruz basin	0–0.4 (0.1)	Shelf	
	0–0.4 (0.1)	Slope	
Bodega basin	0–0.2 (0)	Shelf	
	0–0.1 (0)	Slope	
Point Arena basin	0–0.2 (0)	Shelf	
	0–0.3 (0)	Slope	
Eel River basin	0–1.1 (0.2)	Shelf	
	0–1.5 (0.3)	Slope	
Sacramento basin			0.5–3.3 (1.5)
San Joaquin basin			0.6–5.2 (2.1)
Los Angeles basin			0.2–1.5 (0.6)

TABLE 7.12 *(continued)*

Region	Offshore (shelf 0–650 feet, slope below 650 feet)	Onshore
Ventura basin		0.3–2.4 (1.1)
Central Coastal basin		0–0.2 (0.1)
Humboldt basin		0–0.2 (0.1)
Mean total	(5.2)	(5.7)

a For an explanation of estimates, see Table 7.2.
Source: See Table 7.2.

TABLE 7.13   Estimated Ultimate California Natural Gas Recovery (in trillion cubic feet)

Cumulative production (to the end of 1980)	28.78
Proved reserves	5.01
Inferred reserves (growth)	3.4
Undiscovered recoverable resources	10.9
Total California natural gas	48.09

TABLE 7.14   Estimated Ultimate California Gas Liquids Recovery (in billion barrels)

Cumulative production (to the end of 1980)	.78
Proved reserves	.120
Inferred reserves (growth)	.09
Undiscovered resources	.5
Total California natural gas liquids	1.490

TABLE 7.15   Estimated Remaining California Petroleum

	Oil (billion barrels)	Gas Liquids (billion barrels)	Natural Gas (trillion cubic feet)
Proved reserves	3.787	.120	5.01
Inferred reserves (growth)	2.8	.09	3.4
Undiscovered resources	7.0	.5	10.9
Total remaining petroleum	13.587	.710	19.31

TABLE 7.16   Wyoming Oil Production History (in million barrels)

Year	Production	Reserves	Reserves/Production Ratio
1969	154	997	6:1
1970	155	1,017	7:1
1971	145	997	7:1
1972	136	950	7:1
1973	138	917	7:1
1974	138	903	7:1
1975	131	877	7:1
1976	131	828	6:1
1977	135	816	6:1
1978	135	805	6:1
1979	124	809	7:1
1980	125	893	7:1
Total	1,647		

Reserve additions for the decade totaled 1,389 million barrels.

Source: See Table 7.1.

TABLE 7.17   Colorado Oil Production History (in million barrels)

Year	Production	Reserves	Reserves/Production Ratio
1969	28	401	14:1
1970	25	389	16:1
1971	27	333	12:1
1972	32	326	10:1
1973	37	305	8:1
1974	37	289	8:1
1975	37	276	7:1
1976	39	252	6:1
1977	39	242	6:1
1978	37	198	5:1
1979	32	152	5:1
1980	32	175	5:1
Total	402		

Reserve additions for the decaue totaled 148 million barrels.

Source: See Table 7.1.

TABLE 7.18   North Dakota Oil Production History (in million barrels)

Year	Production	Reserves	Reserves/Production Ratio
1969	23	235	10:1
1970	22	192	9:1
1971	22	174	8:1
1972	21	166	8:1
1973	20	180	9:1
1974	20	173	9:1
1975	20	158	8:1
1976	22	150	7:1
1977	23	150	7:1
1978	25	161	6:1
1979	29	175	6:1
1980	38	177	5:1
Total	285		

Reserve additions for the decade totaled 204 million barrels.

Source: See Table 7.1.

TABLE 7.19   Montana Oil Production History (in million barrels)

Year	Production	Reserves	Reserves/Production Ratio
1969	44	276	6:1
1970	38	242	6:1
1971	35	228	7:1
1972	34	241	7:1
1973	35	219	6:1
1974	35	207	6:1
1975	33	164	5:1
1976	33	153	5:1
1977	33	152	5:1
1978	30	140	5:1
1979	29	136	5:1
1980	29	160	6:1
Total	408		

Reserve additions for the decade totaled 248 million barrels.

Source: See Table 7.1.

TABLE 7.20   Utah Oil Production History (in million barrels)

Year	Production	Reserves	Reserves/Production Ratio
1969	23	195	8:1
1970	23	182	8:1
1971	24	166	7:1
1972	26	244	9:1
1973	33	265	8:1
1974	39	251	6:1
1975	40	208	5:1
1976	34	183	5:1
1977	33	182	5:1
1978	31	155	5:1
1979	27	144	5:1
1980	25	142	6:1
Total	358		

Reserve additions for the decade totaled 282 million barrels.

Source: See Table 7.1.

TABLE 7.21   Western New Mexico Oil Production History (in million barrels)

Year	Production	Reserves	Reserves/Production Ratio
1969	6	29	5:1
1970	6	23	4:1
1971	6	23	4:1
1972	6	24	4:1
1973	5	24	5:1
1974	6	27	4:1
1975	4	25	6:1
1976	4	21	5:1
1977	4	19	5:1
1978	4	17	4:1
1979	4	15	4:1
1980	4	21	5:1
Total	59		

Reserve additions for the decade totaled 45 million barrels.

Source: See Table 7.1.

TABLE 7.22   Rocky Mountain and Northern Great Plains Region Oil
Production History (in million barrels)

Year	Production	Reserves	Reserves/Production Ratio
1969	278	2,133	8:1
1970	269	2,045	8:1
1971	259	1,921	7:1
1972	255	1,951	8:1
1973	268	1,910	7:1
1974	275	1,850	7:1
1975	265	1,708	6:1
1976	263	1,587	6:1
1977	267	1,561	6:1
1978	262	1,476	6:1
1979	245	1,431	6:1
1980	253	1,568	6:1
Total	3,159		

Reserve additions for the decade totaled 2,316 million barrels.

Source: See Table 7.1.

TABLE 7.23   Estimated Undiscovered Recoverable Conventional Oil Resources
of the Rocky Mountain and Northern Great Plains Region (in
million barrels)

Area	Estimate	
Eastern Basin & Range (Idaho, Utah, and Nevada)	200–11,900	(3,300)[a]
Western Basin & Range (Nevada)	0–1,900	(400)
Paradox basin (Utah, Colorado)	400–3,200	(1,200)
Uinta-Piceance-Eagle basins (Utah, Colorado)	400–3,800	(1,600)
Park basins (Colorado)	0–200	(100)
San Juan basin (Colorado, New Mexico)	0–400	(100)
Wyoming-Utah-Idaho-Montana Overthrust Belt	2,700–15,300	(7,300)
Northern Arizona	0–1,000	(300)
Southern Arizona–Southeast New Mexico	0–1,300	(300)
Williston basin (Montana, North Dakota, South Dakota)	400–3,200	(1,400)
Sweetgrass arch (Montana)	100–1,200	(400)
Central Montana	0–400	(200)
Southwestern Montana	0–600	(200)
Wind River basin (Wyoming)	200–1,400	(600)
Powder River basin (Wyoming)	500–2,800	(1,400)
Southwestern Wyoming basins (Green River)	800–6,000	(2,600)

TABLE 7.23 *(continued)*

Area	Estimate	
Big Horn basin (Wyoming)	300–2,300	(1,000)
Denver basin (Colorado, Nebraska)	200–1,800	(800)
Las Animas arch (Colorado)	0–300	(100)
Raton basin–Sierra Grande uplift (New Mexico, Colorado)	0–600	(200)
Mean total		(23,500)

ª For an explanation of estimates, see Table 7.2.
Source: See Table 7.2.

TABLE 7.24   Estimated Ultimate Rocky Mountain and Northern Great Plains Oil Recovery (in billion barrels)

Cumulative production (to the end of 1980)	8.496
Proved reserves	1.568
Inferred reserves (growth)	4.6
Undiscovered resources	23.5
Total Rocky Mountain and northern Great Plains oil	38.164

TABLE 7.25   Wyoming Gas Production History (in trillion cubic feet)

Year	Production	Reserves	Reserves/Production Ratio
1969	.318	3.895	12:1
1970	.341	4.199	12:1
1971	.361	4.084	11:1
1972	.379	4.034	11:1
1973	.377	4.054	11:1
1974	.337	3.865	11:1
1975	.299	3.649	12:1
1976	.317	3.651	12:1
1977	.338	3.909	12:1
1978	.362	4.263	12:1
1979	.392	4.630	12:1
1980	.416	5.598	13:1
Total	4.237		

Reserve additions for the decade totaled 5.622 trillion cubic feet.

Source: See Table 7.1.

TABLE 7.26   Colorado Gas Production History (in trillion cubic feet)

Year	Production	Reserves	Reserves/Production Ratio
1969	.109	1.577	14:1
1970	.104	1.604	15:1
1971	.102	1.795	18:1
1972	.110	1.632	15:1
1973	.134	1.843	14:1
1974	.132	1.857	14:1
1975	.165	1.864	11:1
1976	.178	1.856	10:1
1977	.187	1.989	11:1
1978	.186	1.930	10:1
1979	.191	2.004	10:1
1980	.200	2.299	11:1
Total	1.798		

Reserve additions for the decade totaled 2.411 trillion cubic feet.

Source: See Table 7.1.

TABLE 7.27   North Dakota Gas Production History (in trillion cubic feet)

Year	Production	Reserves	Reserves/Production Ratio
1969	.044	.649	15:1
1970	.039	.567	15:1
1971	.041	.504	12:1
1972	.037	.442	12:1
1973	.037	.448	12:1
1974	.032	.433	14:1
1975	.028	.417	15:1
1976	.024	.406	17:1
1977	.025	.392	16:1
1978	.030	.411	14:1
1979	.038	.489	13:1
1980	.037	.537	15:1
Total	.412		

Reserve additions for the decade totaled 0.256 trillion cubic feet.

Source: See Table 7.1.

TABLE 7.28   Montana Gas Production History (in trillion cubic feet)

Year	Production	Reserves	Reserves/Production Ratio
1969	.044	1.110	25:1
1970	.039	1.100	28:1
1971	.034	1.025	30:1
1972	.031	1.064	34:1
1973	.060	1.092	18:1
1974	.051	.901	18:1
1975	.052	.930	18:1
1976	.051	1.106	22:1
1977	.054	1.044	19:1
1978	.057	.992	17:1
1979	.062	1.036	17:1
1980	.061	1.287	21:1
Total	.596		

Reserve additions for the decade totaled 0.729 trillion cubic feet.

Source: See Table 7.1.

TABLE 7.29   Utah Gas Production History (in trillion cubic feet)

Year	Production	Reserves	Reserves/Production Ratio
1969	.053	1.089	21:1
1970	.045	1.064	24:1
1971	.043	.981	23:1
1972	.046	1.021	22:1
1973	.051	1.023	20:1
1974	.056	1.029	18:1
1975	.059	.914	15:1
1976	.057	.827	15:1
1977	.063	.747	12:1
1978	.060	.696	12:1
1979	.047	.675	14:1
1980	.061	.855	14:1
Total	.641		

Reserve additions for the decade totaled 0.354 trillion cubic feet.

Source: See Table 7.1.

TABLE 7.30   Western New Mexico Gas Production History (in trillion cubic feet)

Year	Production	Reserves	Reserves/Production Ratio
1969	.553	9.075	16:1
1970	.533	8.650	16:1
1971	.564	8.568	15:1
1972	.596	8.161	14:1
1973	.559	7.985	14:1
1974	.555	7.614	14:1
1975	.520	7.773	14:1
1976	.522	7.990	15:1
1977	.531	8.066	15:1
1978	.538	9.646	18:1
1979	.558	9.843	18:1
1980	.549	9.908	18:1
Total	6.578		

Reserve additions for the decade totaled 6.858 trillion cubic feet.

Source: See Table 7.1.

TABLE 7.31   Rocky Mountain and Northern Great Plains Region Gas Production History (in trillion cubic feet)

Year	Production	Reserves	Reserves/Production Ratio
1969	1.121	17.395	16:1
1970	1.101	17.184	16:1
1971	1.145	16.957	15:1
1972	1.199	16.354	14:1
1973	1.218	16.445	14:1
1974	1.163	15.699	13:1
1975	1.123	15.547	14:1
1976	1.149	15.836	14:1
1977	1.198	16.147	13:1
1978	1.233	17.938	15:1
1979	1.288	18.677	15:1
1980	1.324	20.484	15:1
Total	14.262		

Reserve additions for the decade totaled 16.230 trillion cubic feet.

Source: See Table 7.1.

TABLE 7.32   Estimated Undiscovered Recoverable Conventional Gas Resources
of the Rocky Mountain and Northern Great Plains Region
(in trillion cubic feet)

Area	Estimate	
Eastern Basin & Range (Idaho, Utah, Nevada)	0.3–39.3	(10.6)[a]
Western Basin & Range (Nevada)	0–5.4	(1.2)
Paradox basin (Utah, Colorado)	0.7–10.6	(3.8)
Uinta-Piceance-Eagle basins (Utah, Colorado)	2.2–22.4	(8.9)
Park basins (Colorado)	0–0.6	(0.2)
Albuquerque-Sante Fe-San Luis Rift basins (Colorado, New Mexico)	0–1.5	(0.4)
San Juan basin (New Mexico, Colorado)	0.9–7.9	(3.3)
Wyoming-Utah-Idaho-Montana Overthrust Belt	24.1–143.5	(67.7)
Northern Arizona	0–2.2	(0.5)
South-central New Mexico	0–1.5	(0.4)
Southern Arizona–Southwestern New Mexico	0–9.7	(2.4)
Williston basin (Montana, North Dakota, South Dakota)	1.0–7.5	(3.4)
Sweetgrass arch (Montana)	0.6–6.9	(2.8)
Central Montana	0.1–1.8	(0.6)
Southwest Montana	0–1.6	(0.4)
Wind River basin (Wyoming)	0.8–4.9	(2.3)
Powder River basin (Wyoming)	0.9–5.4	(2.5)
Southwest Wyoming basins (Green River)	6.2–40.2	(18.6)
Big Horn basin (Wyoming)	0.8–5.3	(2.4)
Denver basin (Colorado)	0.7–4.9	(2.2)
Las Animas arch (Colorado)	0.3–2.3	(1.0)
Raton basin–Sierre Grande uplift (New Mexico, Colorado)	0–1.7	(0.6)
Mean total		(136.2)

[a] For an explanation of estimates, see Table 7.2.
Source: See Table 7.2.

TABLE 7.33   Estimated Ultimate Rocky Mountain and Northern Great Plains
Natural Gas Recovery (in trillion cubic feet)

Cumulative production (to the end of 1980)	29.450
Proved reserves	20.484
Inferred reserves (growth)	11.1
Undiscovered resources	136.2
Total Rocky Mountain and northern Great Plains gas	197.234

TABLE 7.34   Estimated Ultimate Rocky Mountain and Northern Great Plains
Gas Liquids Recovery (in billion barrels)

Cumulative production (to the end of 1980)	1.178
Proved reserves	1.071
Inferred reserves (growth)	.44
Undiscovered resources	4.0
Total Rocky Mountain and northern Great Plains gas liquids	6.689

TABLE 7.35   Estimated Remaining Rocky Mountain and Northern Great
Plains Petroleum

	Oil *(billion barrels)*	*Gas Liquids* *(billion barrels)*	*Natural Gas* *(trillion cubic feet)*
Proved reserves	1.568	1.071	20.484
Inferred reserves (growth)	4.6	.44	11.1
Undiscovered resources	23.5	4.0	136.2
Total remaining petroleum	29.668	5.511	167.784

TABLE 7.36   West Texas Oil Production History (in million barrels)

Year	*Production*	*Reserves*	*Reserves/Production Ratio*
1969	631	6,661	11:1
1970	675	7,138	11:1
1971	676	7,348	11:1
1972	720	6,995	10:1
1973	737	6,796	9:1
1974	735	6,463	9:1
1975	723	5,933	8:1
1976	706	5,434	8:1
1977	685	5,055	7:1
1978	657	4,635	7:1
1979	636	4,936	8:1
1980	607	4,980	8:1
Total	8,188		

Reserve additions for the decade totaled 5,876 million barrels.

Source: See Table 7.1.

TABLE 7.37   Eastern New Mexico Oil Production History (in million barrels)

Year	Production	Reserves	Reserves/Production Ratio
1969	116	811	7:1
1970	117	738	6:1
1971	108	634	6:1
1972	100	558	6:1
1973	91	619	7:1
1974	88	598	7:1
1975	86	563	7:1
1976	84	514	6:1
1977	79	472	6:1
1978	75	496	6:1
1979	70	447	6:1
1980	63	427	7:1
Total	1,077		

Reserve additions for the decade totaled 577 million barrels.

Source: See Table 7.1.

TABLE 7.38   West Texas and Eastern New Mexico Region Oil Production
History (in million barrels)

Year	Production	Reserves	Reserves/Production Ratio
1969	747	7,472	10:1
1970	792	7,876	10:1
1971	784	7,982	10:1
1972	820	7,553	9:1
1973	828	7,415	9:1
1974	823	7,061	9:1
1975	809	6,496	8:1
1976	790	5,948	8:1
1977	764	5,527	7:1
1978	732	5,104	7:1
1979	706	5,383	8:1
1980	670	5,407	8:1
Total	9,265		

Reserve additions for the decade totaled 6,453 million barrels.

Source: See Table 7.1.

TABLE 7.39   Estimated Undiscovered Recoverable Conventional Oil
Resources of the West Texas and Eastern New Mexico Region
(in million barrels)

Area	Estimate	
Permian basin	1,000–6,300	(2,900)[a]
Bend arch–Fort Worth basin	700–4,200	(2,000)
Marathon fold belt	0–1,700	(500)
Mean total		(5,400)

[a] For an explanation of estimates, see Table 7.2.
Source: See Table 7.2.

TABLE 7.40   Estimated Ultimate West Texas and Eastern New Mexico Oil
Recovery (in billion barrels)

Cumulative production (to end of 1980)	25.857
Proved reserves	5.407
Inferred reserves (growth)	5.3
Undiscovered resources	5.4
Total West Texas and eastern New Mexico oil	41.964

TABLE 7.41   West Texas Gas Production History (in trillion cubic feet)

Year	Production	Reserves	Reserves/Production Ratio
1969	2.018	28.153	14:1
1970	2.383	27.352	11:1
1971	2.485	23.759	10:1
1972	2.720	22.611	8:1
1973	2.732	23.096	8:1
1974	2.732	20.257	7:1
1975	2.349	17.916	8:1
1976	2.176	16.186	7:1
1977	2.082	14.522	7:1
1978	1.956	12.858	7:1
1979	1.850	12.349	7:1
1980	1.775	11.187	6:1
Total	27.258		

Reserve additions for the decade totaled 8.274 trillion cubic feet.

Source: See Table 7.1.

TABLE 7.42   Eastern New Mexico Gas Production History (in trillion cubic feet)

Year	Production	Reserves	Reserves/Production Ratio
1969	.535	5.197	10:1
1970	.569	4.635	8:1
1971	.559	4.497	8:1
1972	.578	4.175	7:1
1973	.635	4.487	7:1
1974	.646	4.303	7:1
1975	.601	3.957	7:1
1976	.697	3.897	6:1
1977	.611	3.840	6:1
1978	.531	3.592	7:1
1979	.555	3.597	6:1
1980	.515	3.150	6:1
Total	7.032		

Reserve additions for the decade totaled 4.450 trillion cubic feet.

Source: See Table 7.1.

TABLE 7.43   West Texas and Eastern New Mexico Region Gas Production History (in trillion cubic feet)

Year	Production	Reserves	Reserves/Production Ratio
1969	2.553	33.350	13:1
1970	2.952	31.987	11:1
1971	3.044	28.256	9:1
1972	3.298	26.786	8:1
1973	3.367	27.583	8:1
1974	3.378	24.560	7:1
1975	2.950	21.873	7:1
1976	2.873	20.083	7:1
1977	2.693	18.362	7:1
1978	2.487	16.450	7:1
1979	2.405	15.946	7:1
1980	2.290	14.337	6:1
Total	34.290		

Reserve additions for the decade totaled 12.724 trillion cubic feet.

Source: See Table 7.1.

TABLE 7.44   Estimated Undiscovered Recoverable Conventional Natural Gas
Resources of the West Texas and Eastern New Mexico Region
(in trillion cubic feet)

Area	Estimate	
Permian basin	11.1–73.0	(33.4)[a]
Bend arch–Fort Worth basin	3.2–16.5	(8.2)
Marathon fold belt	0–4.2	(1.1)
Mean total		(42.7)

[a]For an explanation of estimates, see Table 7.2.
Source: See Table 7.2.

TABLE 7.45   Estimated Ultimate West Texas and Eastern New Mexico Natural
Gas Recovery (in trillion cubic feet)

Cumulative production (to end of 1980)	74.384
Proved reserves	14.337
Inferred reserves (growth)	16.3
Undiscovered resources	42.7
Total West Texas and eastern New Mexico gas	147.721

TABLE 7.46   Estimated Ultimate West Texas and Eastern New Mexico Gas
Liquids Recovery (in billion barrels)

Cumulative production (to end of 1980)	5.579
Proved reserves	1.309
Inferred reserves (growth)	1.22
Undiscovered resources	1.5
Total West Texas and eastern New Mexico gas liquids	9.608

TABLE 7.47   Estimated Remaining West Texas and Eastern New Mexico
Petroleum

	Oil (billion barrels)	Gas Liquids (billion barrels)	Natural Gas (trillion cubic feet)
Proved reserves	5.407	1.309	14.337
Inferred reserves (growth)	5.3	1.22	16.3
Undiscovered resources	5.4	1.5	42.7
Total remaining petroleum	16.107	4.029	73.337

TABLE 7.48   Texas Gulf Coast Oil Production History (in million barrels)

Year	Production	Reserves	Reserves/Production Ratio
1969	446	6,166	14:1
1970	504	5,839	12:1
1971	482	5,476	11:1
1972	520	4,971	10:1
1973	499	4,789	10:1
1974	471	4,370	9:1
1975	446	3,992	9:1
1976	427	3,646	9:1
1977	400	3,279	8:1
1978	367	2,928	8:1
1979	329	2,548	8:1
1980	315	2,425	8:1
Total	5,206		

Reserve additions for the decade totaled 1,019 million barrels.

Source: See Table 7.1.

TABLE 7.49   Louisiana Oil Production History (in million barrels)

Year	Production	Reserves	Reserves/Production Ratio
1969	727	5,689	8:1
1970	785	5,710	7:1
1971	833	5,399	6:1
1972	772	5,028	7:1
1973	710	4,577	6:1
1974	627	4,227	7:1
1975	557	3,827	7:1
1976	525	3,471	7:1
1977	486	3,113	6:1
1978	455	2,893	6:1
1979	413	2,555	6:1
1980	400	2,528	6:1
Total	7,290		

Reserve additions for the decade totaled 3,402 million barrels.

Source: See Table 7.1.

TABLE 7.50   Mississippi Oil Production History (in million barrels)

Year	Production	Reserves	Reserves/Production Ratio
1969	64	360	6:1
1970	65	355	5:1
1971	64	342	5:1
1972	58	312	5:1
1973	56	291	5:1
1974	49	261	5:1
1975	47	231	5:1
1976	45	214	5:1
1977	42	203	5:1
1978	38	188	5:1
1979	35	169	5:1
1980	30	143	5:1
Total	593		

Reserve additions for the decade totaled 312 million barrels.

Source: See Table 7.1.

TABLE 7.51   Alabama Oil Production History (in million barrels)

Year	Production	Reserves	Reserves/Production Ratio
1969	7	67	10:1
1970	7	65	9:1
1971	8	61	8:1
1972	10	57	6:1
1973	11	54	5:1
1974	11	69	6:1
1975	10	61	6:1
1976	12	53	4:1
1977	13	44	3:1
1978	11	33	3:1
1979	10	28	3:1
1980	9	34	4:1
Total	119		

Reserve additions for the decade totaled 79 million barrels.

Source: See Table 7.1.

TABLE 7.52   Arkansas Oil Production History (in million barrels)

Year	Production	Reserves	Reserves/Production Ratio
1969	17	127	7:1
1970	17	130	8:1
1971	18	118	7:1
1972	18	113	6:1
1973	18	106	6:1
1974	16	106	7:1
1975	16	96	6:1
1976	18	93	5:1
1977	20	86	4:1
1978	20	94	5:1
1979	18	107	6:1
1980	16	107	7:1
Total	212		

Reserve additions for the decade totaled 175 million barrels.

Source: See Table 7.1.

TABLE 7.53   Florida Oil Production History (in million barrels)

Year	Production	Reserves	Reserves/Production Ratio
1969	2	9	4:1
1970	3	75	25:1
1971	5	204	41:1
1972	17	208	12:1
1973	33	184	6:1
1974	36	303	8:1
1975	42	263	6:1
1976	44	250	6:1
1977	42	209	5:1
1978	48	169	4:1
1979	48	123	3:1
1980	42	129	3:1
Total	362		

Reserve additions for the decade totaled 480 million barrels.

Source: See Table 7.1.

TABLE 7.54   Gulf Coast Region Oil Production History (in million barrels)

Year	Production	Reserves	Reserves/Production Ratio
1969	1,263	12,418	10:1
1970	1,381	12,174	9:1
1971	1,410	11,600	8:1
1972	1,395	10,689	8:1
1973	1,327	10,001	8:1
1974	1,210	9,336	8:1
1975	1,118	8,470	8:1
1976	1,071	7,727	7:1
1977	1,003	6,934	7:1
1978	939	6,308	7:1
1979	853	5,530	6:1
1980	812	5,366	7:1
Total	13,782		

Reserve additions for the decade totaled 5,467 million barrels.

Source: See Table 7.1.

TABLE 7.55   Estimated Undiscovered Recoverable Conventional Oil Resources of the Gulf Coast Region (in million barrels)

Area	Offshore (shelf 0–650 feet, slope below 650 feet)			Onshore	
Eastern Gulf	0–3,800	(1,200)[a]	Shelf		
	0–1,000	(200)	Slope		
Western Gulf	1,100–5,700	(2,800)	Shelf	1,200–8,000	(3,600)
	900–4,800	(2,400)	Slope		
East Texas basin				300–2,700	(1,200)
Louisiana-Mississippi salt basins				800–5,000	(2,300)
Black Warrior basin				0–500	(200)
Mean total		(6,600)			(7,300)

[a] For an explanation of estimates, see Table 7.2.
Source: See Table 7.2.

TABLE 7.56   Estimated Ultimate Gulf Coast Oil Recovery (in billion barrels)

Cumulative production (to the end of 1980)	41.312
Proved reserves	5.366
Inferred reserves (growth)	6.3
Undiscovered resources	13.9
Total Gulf Coast oil	66.878

TABLE 7.57   Texas Gulf Coast Gas Production History (in trillion cubic feet)

Year	Production	Reserves	Reserves/Production Ratio
1969	4.520	74.920	17:1
1970	4.559	71.293	16:1
1971	4.473	68.502	15:1
1972	4.354	62.932	14:1
1973	4.272	54.350	13:1
1974	3.958	49.084	12:1
1975	3.526	43.488	12:1
1976	3.580	39.990	11:1
1977	3.611	39.810	11:1
1978	3.564	33.976	10:1
1979	3.963	31.853	8:1
1980	3.863	30.888	8:1
Total	48.243		

Reserve additions for the decade totaled –0.309 trillion cubic feet. (Due to extensive revisions, reserve additions are statistically negative.)

Source: See Table 7.1.

TABLE 7.58   Louisiana Gas Production History (in trillion cubic feet)

Year	Production	Reserves	Reserves/Production Ratio
1969	7.278	84.944	12:1
1970	7.790	82.819	11:1
1971	8.114	78.453	10:1
1972	8.412	74.978	9:1
1973	8.458	68.973	8:1
1974	7.714	63.878	8:1
1975	7.181	61.098	9:1
1976	6.975	57.251	8:1
1977	7.029	52.379	7:1
1978	7.062	49.353	7:1
1979	7.224	45.237	6:1
1980	6.744	42.781	6:1
Total	89.981		

Reserve additions for the decade totaled 40.540 trillion cubic feet.

Source: See Table 7.1.

TABLE 7.59   Mississippi Gas Production History (in trillion cubic feet)

Year	Production	Reserves	Reserves/Production Ratio
1969	.166	1.406	8:1
1970	.156	1.325	8:1
1971	.132	1.107	8:1
1972	.110	1.003	9:1
1973	.097	1.089	11:1
1974	.084	.987	12:1
1975	.080	1.127	14:1
1976	.080	.991	12:1
1977	.089	1.221	14:1
1978	.132	1.320	10:1
1979	.160	1.345	8:1
1980	.170	1.582	9:1
Total	1.456		

Reserve additions for the decade totaled 1.466 trillion cubic feet.

Source: See Table 7.1.

TABLE 7.60   Alabama Gas Production History (in trillion cubic feet)

Year	Production	Reserves	Reserves/Production Ratio
1969	.000	.057	
1970	.001	.162	162:1
1971	.001	.181	181:1
1972	.005	.246	49:1
1973	.008	.327	41:1
1974	.017	.507	30:1
1975	.021	.771	37:1
1976	.024	.707	29:1
1977	.038	.746	20:1
1978	.056	.751	13:1
1979	.059	.933	16:1
1980	.064	.910	14:1
Total	.294		

Reserve additions for the decade totaled 1.147 trillion cubic feet.

Source: See Table 7.1.

TABLE 7.61   Arkansas Gas Production History (in trillion cubic feet)

Year	Production	Reserves	Reserves/Production Ratio
1969	.166	2.596	16:1
1970	.184	2.545	14:1
1971	.174	2.396	14:1
1972	.169	2.422	14:1
1973	.164	2.234	14:1
1974	.124	2.089	17:1
1975	.118	1.969	17:1
1976	.120	1.703	14:1
1977	.107	1.655	15:1
1978	.106	1.601	15:1
1979	.130	1.606	12:1
1980	.117	1.673	14:1
Total	1.679		

Reserve additions for the decade totaled 0.590 trillion cubic feet.

Source: See Table 7.1.

TABLE 7.62   Gulf Coast Region Gas Production History (in trillion cubic feet)

Year	Production	Reserves	Reserves/Production Ratio
1969	12.130	163.923	14:1
1970	12.690	158.144	12:1
1971	12.894	147.151	11:1
1972	13.050	141.581	11:1
1973	12.999	126.973	10:1
1974	11.897	116.545	10:1
1975	10.926	108.453	10:1
1976	10.779	100.642	10:1
1977	10.874	95.811	9:1
1978	10.920	87.001	8:1
1979	11.536	80.974	7:1
1980	10.958	77.834	7:1
Total	141.653		

Reserve additions for the decade totaled 43.434 trillion cubic feet.

Source: See Table 7.1.

TABLE 7.63 Estimated Undiscovered Recoverable Conventional Natural Gas Resources of the Gulf Coast Region (in trillion cubic feet)

Area	Offshore (shelf 0–650 feet, slope below 650 feet)		Onshore
Eastern Gulf	0–9.0	(2.4)[a] Shelf	
	0–2.6	(0.5) Slope	
Western Gulf	18.1–82.2	(42.9) Shelf	30.8–229.3 (101.2)
	8.8–56.3	(26.1) Slope	
East Texas basin			2.4–15.6 (7.2)
Louisiana-Mississippi salt basins			4.1–38.9 (16.0)
Black Warrior basin			0.4–3.4 (1.4)
Mean total		(71.9)	(125.8)

[a] For an explanation of estimates, see Table 7.2.
Source: See Table 7.2.

TABLE 7.64 Estimated Ultimate Gulf Coast Natural Gas Recovery (in trillion cubic feet)

Cumulative production (to the end of 1980)	296.030
Proved reserves	77.834
Inferred reserves (growth)	111.1
Undiscovered resources	197.7
Total Gulf Coast gas	682.664

TABLE 7.65 Estimated Ultimate Gulf Coast Gas Liquids Recovery (in billion barrels)

Cumulative production (to the end of 1980)	8.644
Proved reserves	2.591
Inferred reserves (growth)	3.2
Undiscovered resources	5.6
Total Gulf Coast gas liquids	20.035

TABLE 7.66   Estimated Remaining Gulf Coast Petroleum

	Oil *(billion barrels)*	*Gas Liquids* *(billion barrels)*	*Natural Gas* *(trillion cubic feet)*
Proved reserves	5.366	2.591	77.834
Inferred reserves (growth)	6.3	3.2	111.1
Undiscovered resources	13.9	5.6	197.7
Total remaining petroleum	25.566	11.391	386.634

TABLE 7.67   Texas Panhandle Oil Production History (in million barrels)

Year	*Production*	*Reserves*	*Reserves/Production Ratio*
1969	29	235	8:1
1970	27	218	8:1
1971	25	200	8:1
1972	23	177	8:1
1973	20	170	8:1
1974	19	169	9:1
1975	18	155	9:1
1976	18	147	8:1
1977	17	133	8:1
1978	16	135	8:1
1979	16	152	10:1
1980	15	139	9:1
Total	243		

Reserve additions during the decade totaled 118 million barrels.

Source: See Table 7.1.

TABLE 7.68   Oklahoma Oil Production History (in million barrels)

Year	Production	Reserves	Reserves/Production Ratio
1969	215	1,390	6:1
1970	213	1,351	6:1
1971	210	1,405	7:1
1972	198	1,303	7:1
1973	181	1,271	7:1
1974	166	1,232	7:1
1975	154	1,240	8:1
1976	152	1,187	8:1
1977	144	1,121	8:1
1978	137	1,073	8:1
1979	128	1,005	8:1
1980	131	922	7:1
Total	2,029		

Reserve additions during the decade totaled 1,346 million barrels.

Source: See Table 7.1.

TABLE 7.69   Kansas Oil Production History (in million barrels)

Year	Production	Reserves	Reserves/Production Ratio
1969	88	566	6:1
1970	84	539	6:1
1971	78	502	6:1
1972	73	453	6:1
1973	66	401	6:1
1974	62	395	6:1
1975	59	364	6:1
1976	59	362	6:1
1977	57	359	6:1
1978	56	350	6:1
1979	53	345	7:1
1980	54	310	6:1
Total	789		

Reserve additions during the decade totaled 445 million barrels.

Source: See Table 7.1.

TABLE 7.70   Midcontinent Region Oil Production History (in million barrels)

Year	Production	Reserves	Reserves/Production Ratio
1969	332	2,191	7:1
1970	324	2,108	7:1
1971	313	2,107	7:1
1972	294	1,933	7:1
1973	267	1,842	7:1
1974	247	1,796	7:1
1975	231	1,759	8:1
1976	229	1,696	7:1
1977	218	1,613	7:1
1978	209	1,558	7:1
1979	197	1,502	8:1
1980	200	1,371	7:1
Total	3,061		

Reserve additions for the decade totaled 1,909 million barrels.

Source: See Table 7.1.

TABLE 7.71   Estimated Undiscovered Recoverable Conventional Oil Resources of the Midcontinent Region (in million barrels)

Area	Estimate	
Anadarko basin	700–5,100	(2,200)[a]
Arkoma basin	0–1,000	(300)
Central Kansas uplift	100–1,000	(400)
Cherokee platform	0–1,000	(300)
Forest City basin	0–100	(0)
Nemaha ridge	0–200	(100)
Sedgwick basin	100–600	(200)
Southern Oklahoma	200–1,800	(700)
Mean total		(4,200)

[a] For an explanation of estimates, see Table 7.2.
Source: See Table 7.2.

TABLE 7.72   Estimated Ultimate Midcontinent Oil Recovery
             (in billion barrels)

Cumulative production (to end of 1980)	18.487
Proved reserves	1.371
Inferred reserves (growth)	1.5
Undiscovered resources	4.2
Total Midcontinent oil	25.558

TABLE 7.73   Texas Panhandle Gas Production History (in trillion cubic feet)

Year	Production	Reserves	Reserves/Production Ratio
1969	1.121	10.731	10:1
1970	1.279	10.093	8:1
1971	1.209	9.825	8:1
1972	1.196	9.359	8:1
1973	1.237	8.986	7:1
1974	1.253	9.055	7:1
1975	1.168	8.491	7:1
1976	1.141	8.307	7:1
1977	1.133	7.607	7:1
1978	1.010	7.511	7:1
1979	1.000	7.145	7:1
1980	.953	6.777	7:1
Total	13.700		

Reserve additions for the decade totaled 8.625 trillion cubic feet.

Source: See Table 7.1.

TABLE 7.74   Oklahoma Gas Production History (in trillion cubic feet)

Year	Production	Reserves	Reserves/Production Ratio
1969	1.680	17.386	10:1
1970	1.751	16.733	10:1
1971	1.665	15.474	9:1
1972	1.756	14.262	8:1
1973	1.778	13.859	8:1
1974	1.705	13.176	8:1
1975	1.672	12.863	8:1
1976	1.654	12.217	7:1
1977	1.663	11.464	7:1
1978	1.647	11.205	7:1
1979	1.660	10.929	7:1
1980	1.526	10.393	7:1
Total	20.157		

Reserve additions for the decade totaled 11.484 trillion cubic feet.

Source: See Table 7.1.

TABLE 7.75   Kansas Gas Production History (in trillion cubic feet)

Year	Production	Reserves	Reserves/Production Ratio
1969	.886	14.033	16:1
1970	.901	13.228	15:1
1971	.890	12.441	14:1
1972	.866	11.849	14:1
1973	.899	11.615	13:1
1974	.892	11.606	13:1
1975	.847	12.556	15:1
1976	.825	11.846	14:1
1977	.771	11.813	15:1
1978	.844	12.175	14:1
1979	.791	12.400	16:1
1980	.682	11.510	17:1
Total	10.094		

Reserve additions for the decade totaled 6.685 trillion cubic feet.

Source: See Table 7.1.

TABLE 7.76 Midcontinent Region Gas Production History (in trillion cubic feet)

Year	Production	Reserves	Reserves/Production Ratio
1969	3.687	42.150	11:1
1970	3.931	40.054	10:1
1971	3.764	37.740	10:1
1972	3.818	35.470	9:1
1973	3.914	34.460	9:1
1974	3.850	33.837	9:1
1975	3.687	33.910	9:1
1976	3.620	32.370	9:1
1977	3.567	30.884	9:1
1978	3.501	30.891	9:1
1979	3.451	30.474	9:1
1980	3.161	28.680	9:1
Total	43.951		

Reserve additions for the decade totaled 26.794 trillion cubic feet.

Source: See Table 7.1.

TABLE 7.77 Estimated Undiscovered Recoverable Conventional Natural Gas Resources of the Midcontinent Region (in trillion cubic feet)

Area	Estimate	
Anadarko basin	11.8–75.0	(34.7)[a]
Arkoma basin	0.6–10.4	(3.5)
Central Kansas uplift	0.3–1.6	(0.7)
Cherokee platform	0.1–3.4	(1.1)
Forest City basin	0.0–0.1	(0)
Nemaha ridge	0.0–0.7	(0.2)
Sedgwick basin	0.2–1.7	(0.8)
Southern Oklahoma	0.8–8.8	(3.4)
Mean total		(44.4)

[a] See Table 7.2 for an explanation of estimates.
Source: See Table 7.2.

TABLE 7.78   Estimated Ultimate Midcontinent Natural Gas Recovery
(in trillion cubic feet)

Cumulative production (to the end of 1980)	128.687
Proved reserves	28.680
Inferred reserves (growth)	16.9
Undiscovered resources	44.4
Total Midcontinent gas	218.667

TABLE 7.79   Estimated Ultimate Midcontinent Gas Liquids Recovery
(in billion barrels)

Cumulative production (to end of 1980)	4.929
Proved reserves	1.067
Inferred reserves (growth)	.65
Undiscovered resources	1.3
Total Midcontinent gas liquids	7.946

TABLE 7.80   Estimated Remaining Midcontinent Petroleum

	Oil (billion barrels)	Gas Liquids (billion barrels)	Natural Gas (trillion cubic feet)
Proved reserves	1.371	1.067	28.680
Inferred reserves (growth)	1.5	.65	16.9
Undiscovered resources	4.2	1.3	44.4
Total remaining petroleum	7.071	3.017	89.980

TABLE 7.81   Illinois Oil Production History (in million barrels)

Year	Production	Reserves	Reserves/Production Ratio
1969	51	272	5:1
1970	44	229	5:1
1971	39	209	5:1
1972	35	175	5:1
1973	31	152	5:1
1974	28	160	6:1
1975	26	161	6:1
1976	26	155	6:1
1977	26	150	6:1
1978	23	138	6:1
1979	20	133	7:1
1980	17	113	7:1
Total	366		

Reserve additions during the decade totaled 156 million barrels.

Source: See Table 7.1.

TABLE 7.82   Indiana Oil Production History (in million barrels)

Year	Production	Reserves	Reserves/Production Ratio
1969	8	41	5:1
1970	7	37	5:1
1971	7	31	4:1
1972	6	29	5:1
1973	5	27	5:1
1974	5	24	5:1
1975	5	22	4:1
1976	5	22	4:1
1977	5	26	5:1
1978	5	27	5:1
1979	5	25	5:1
1980	4	23	6:1
Total	67		

Reserve additions for the decade totaled 41 million barrels.

Source: See Table 7.1.

TABLE 7.83   Kentucky Oil Production History (in million barrels)

Year	Production	Reserves	Reserves/Production Ratio
1969	13	73	6:1
1970	12	61	5:1
1971	11	53	5:1
1972	10	48	5:1
1973	9	40	4:1
1974	8	37	5:1
1975	8	39	5:1
1976	7	38	5:1
1977	7	35	5:1
1978	6	32	5:1
1979	6	31	5:1
1980	7	35	5:1
Total	104		

Reserve additions for the decade totaled 53 million barrels.

Source: See Table 7.1.

TABLE 7.84   Eastern Interior Region Oil Production History
(in million barrels)

Year	Production	Reserves	Reserves/Production Ratio
1969	72	386	5:1
1970	63	327	5:1
1971	57	293	5:1
1972	51	252	5:1
1973	45	219	5:1
1974	41	221	5:1
1975	39	222	6:1
1976	38	215	6:1
1977	38	211	6:1
1978	34	197	6:1
1979	31	189	6:1
1980	28	171	6:1
Total	537		

Reserve additions for the decade totaled 250 million barrels.

Source: See Table 7.1.

TABLE 7.85   Estimated Undiscovered Recoverable Conventional Oil
              Resources of the Eastern Interior Region (in million barrels)

Area	Estimate	
Illinois basin	100–1,400	(600)[a]
Cincinnati arch	0–500	(200)
Mean total		(800)

[a] See Table 7.2 for an explanation of estimates.
Source: See Table 7.2.

TABLE 7.86   Estimated Ultimate Eastern Interior Oil Recovery
              (in billion barrels)

Cumulative production (to end of 1980)	4.554
Proved reserves	.171
Inferred reserves (growth)	.086
Undiscovered resources	.800
Total Eastern Interior oil	5.611

TABLE 7.87   Michigan Basin Oil Production History (in million barrels)

Year	Production	Reserves	Reserves/Production Ratio
1969	12	52	4:1
1970	12	46	4:1
1971	12	59	5:1
1972	13	62	5:1
1973	14	72	5:1
1974	18	82	5:1
1975	24	93	4:1
1976	30	139	5:1
1977	33	133	4:1
1978	35	190	5:1
1979	35	176	5:1
1980	37	205	6:1
Total	275		

Reserve additions for the decade totaled 416 million barrels.

Source: See Table 7.1.

TABLE 7.88   Estimated Ultimate Michigan Basin Oil Recovery
              (in billion barrels)

Cumulative production (to end of 1980)	.839
Proved reserves	.205
Inferred reserves (growth)	.82
Undiscovered resources	1.1
Total Michigan basin oil	2.964

TABLE 7.89   Michigan Basin Gas Production History (in trillion cubic feet)

Year	Production	Reserves	Reserves/Production Ratio
1969	.034	.155	5:1
1970	.040	.324	8:1
1971	.030	.364	12:1
1972	.039	.680	17:1
1973	.048	.903	19:1
1974	.075	.933	12:1
1975	.106	1.020	10:1
1976	.127	1.053	8:1
1977	.131	1.202	9:1
1978	.159	1.170	7:1
1979	.163	1.138	7:1
1980	.160	1.329	8:1
Total	1.112		

Reserve additions for the decade totaled 2.252 trillion cubic feet.

Source: See Table 7.1.

TABLE 7.90   Estimated Ultimate Michigan Basin Natural Gas Recovery
              (in trillion cubic feet)

Cumulative production (to end of 1980)	1.404
Proved reserves	1.329
Inferred reserves (growth)	1.3
Undiscovered resources	5.1
Total Michigan basin gas	9.133

TABLE 7.91   Estimated Ultimate Michigan Basin Gas Liquids Recovery
(in billion barrels)

Cumulative production (to end of 1980)	.123
Proved reserves	.112
Inferred reserves (growth)	.114
Undiscovered resources	.1
Total Michigan basin gas liquids	.449

TABLE 7.92   Estimated Remaining Michigan Basin Petroleum

	Oil (billion barrels)	Gas Liquids (billion barrels)	Natural Gas (trillion cubic feet)
Proved reserves	.205	.112	1.329
Inferred reserves (growth)	.82	.114	1.3
Undiscovered resources	1.1	.1	5.1
Total remaining petroleum	2.125	.326	7.729

TABLE 7.93   Pennsylvania Oil Production History (in million barrels)

Year	Production	Reserves	Reserves/Production Ratio
1969	4	55	14:1
1970	4	51	13:1
1971	4	47	12:1
1972	3	37	12:1
1973	3	40	13:1
1974	3	50	17:1
1975	3	48	16:1
1976	3	51	17:1
1977	3	49	16:1
1978	3	48	16:1
1979	3	51	17:1
1980	2	54	27:1
Total	38		

Reserve additions during the decade totaled 33 million barrels.

Source: See Table 7.1.

TABLE 7.94   West Virginia Oil Production History (in million barrels)

Year	Production	Reserves	Reserves/Production Ratio
1969	3	53	18:1
1970	3	53	18:1
1971	3	52	17:1
1972	3	34	11:1
1973	2	32	16:1
1974	3	32	11:1
1975	3	31	10:1
1976	3	30	10:1
1977	3	29	10:1
1978	2	30	15:1
1979	2	28	14:1
1980	2	30	15:1
Total	32		

Reserve additions during the decade totaled 6 million barrels.

Source: See Table 7.1.

TABLE 7.95   Ohio Oil Production History (in million barrels)

Year	Production	Reserves	Reserves/Production Ratio
1969	11	127	12:1
1970	10	128	13:1
1971	8	129	16:1
1972	9	127	14:1
1973	8	125	16:1
1974	9	124	14:1
1975	10	121	12:1
1976	10	125	12:1
1977	10	129	13:1
1978	11	131	12:1
1979	12	138	12:1
1980	12	116	10:1
Total	120		

Reserve additions for the decade totaled 98 million barrels.

Source: See Table 7.1.

TABLE 7.96   Appalachian Region Oil Production History (in million barrels)

Year	Production	Reserves	Reserves/Production Ratio
1969	18	235	13:1
1970	17	232	14:1
1971	15	228	15:1
1972	15	198	13:1
1973	13	197	15:1
1974	15	206	14:1
1975	16	200	12:1
1976	16	206	13:1
1977	16	207	13:1
1978	16	209	13:1
1979	17	217	13:1
1980	16	200	12:1
Total	190		

Reserve additions for the decade totaled 137 million barrels.

Source: See Table 7.1.

TABLE 7.97   Estimated Ultimate Appalachian Oil Recovery
(in billion barrels)

Cumulative production (to end of 1980)	2.816
Proved reserves	.200
Inferred reserves (growth)	.1
Undiscovered resources	.6
Total Appalachian oil	3.716

TABLE 7.98   Pennsylvania Gas Production History (in trillion cubic feet)

Year	Production	Reserves	Reserves/Production Ratio
1969	.079	.806	10:1
1970	.078	.792	10:1
1971	.076	.780	10:1
1972	.074	.800	11:1
1973	.079	.889	11:1
1974	.083	.946	11:1
1975	.085	1.085	13:1
1976	.090	1.138	13:1
1977	.092	1.328	14:1
1978	.098	1.511	15:1
1979	.096	1.618	17:1
1980	.068	1.016	15:1
Total	.998		

Reserve additions for the decade totaled 1.129 trillion cubic feet.

Source: See Table 7.1.

TABLE 7.99   West Virginia Gas Production History (in trillion cubic feet)

Year	Production	Reserves	Reserves/Production Ratio
1969	.214	2.109	10:1
1970	.232	2.060	9:1
1971	.218	2.019	9:1
1972	.188	1.971	10:1
1973	.168	1.964	12:1
1974	.160	1.921	12:1
1975	.148	1.939	13:1
1976	.151	1.960	13:1
1977	.147	2.001	14:1
1978	.147	2.302	16:1
1979	.153	2.466	16:1
1980	.151	2.422	16:1
Total	2.077		

Reserve additions for the decade totaled 2.176 trillion cubic feet.

Source: See Table 7.1.

TABLE 7.100   Ohio Gas Production History (in trillion cubic feet)

Year	Production	Reserves	Reserves/Production Ratio
1969	.051	.392	8:1
1970	.074	.557	8:1
1971	.083	.616	7:1
1972	.090	.714	8:1
1973	.090	.808	9:1
1974	.094	.960	10:1
1975	.086	.969	11:1
1976	.090	1.026	11:1
1977	.100	1.093	11:1
1978	.115	1.177	10:1
1979	.125	1.313	11:1
1980	.137	1.508	11:1
Total	1.135		

Reserve additions for the decade totaled 2.200 trillion cubic feet.

Source: See Table 7.1.

TABLE 7.101   New York Gas Production History (in trillion cubic feet)

Year	Production	Reserves	Reserves/Production Ratio
1969	.005	.028	6:1
1970	.003	.023	8:1
1971	.002	.023	12:1
1972	.003	.032	11:1
1973	.004	.032	8:1
1974	.004	.062	16:1
1975	.007	.103	15:1
1976	.009	.134	15:1
1977	.010	.138	14:1
1978	.013	.150	12:1
1979	.015	.163	11:1
1980	.014	.161	12:1
Total	.089		

Reserve additions for the decade totaled 0.217 trillion cubic feet.

Source: See Table 7.1.

TABLE 7.102   Kentucky Gas Production History (in trillion cubic feet)

Year	Production	Reserves	Reserves/Production Ratio
1969	.079	.870	11:1
1970	.067	.854	13:1
1971	.065	.819	13:1
1972	.064	.793	12:1
1973	.062	.752	12:1
1974	.060	.727	12:1
1975	.060	.681	11:1
1976	.059	.645	11:1
1977	.057	.609	11:1
1978	.057	.584	10:1
1979	.056	.557	10:1
1980	.060	.605	10:1
Total	.746		

Reserve additions for the decade totaled 0.402 trillion cubic feet.

Source: See Table 7.1.

TABLE 7.103   Appalachian Region Gas Production History
(in trillion cubic feet)

Year	Production	Reserves	Reserves/Production Ratio
1969	.428	4.205	10:1
1970	.454	4.286	9:1
1971	.444	4.257	10:1
1972	.419	4.310	10:1
1973	.403	4.445	11:1
1974	.401	4.616	12:1
1975	.386	4.777	12:1
1976	.399	4.903	12:1
1977	.406	5.169	13:1
1978	.430	5.724	13:1
1979	.445	6.117	14:1
1980	.430	5.712	13:1
Total	5.045		

Reserve additions for the decade totaled 6.124 trillion cubic feet.

Source: See Table 7.1.

TABLE 7.104   Estimated Undiscovered Recoverable Conventional Natural
Gas Resources of the Appalachian Region (in trillion cubic feet)

Area	Estimate	
Appalachian basin	5.5–44.4	(19.1)[a]
Blue Ridge overthrust belt	0–5.0	(1.1)
Mean total		(20.2)

[a] See Table 7.2 for an explanation of estimates.
Source: See Table 7.2.

TABLE 7.105   Estimated Ultimate Appalachian Natural Gas Recovery
(in trillion cubic feet)

Cumulative production (to end of 1980)	34.630
Proved reserves	5.712
Inferred reserves (growth)	5.4
Undiscovered resources	20.2
Total Appalachian gas	65.942

TABLE 7.106   Estimated Ultimate Appalachian Gas Liquids Recovery
(in billion barrels)

Cumulative production (to end of 1980)	.727
Proved reserves	.122
Inferred reserves (growth)	.11
Undiscovered resources	.2
Total Appalachian gas liquids	1.159

TABLE 7.107   Estimated Remaining Appalachian Petroleum

	Oil (billion barrels)	Gas Liquids (billion barrels)	Natural Gas (trillion cubic feet)
Proved reserves	.200	.122	5.712
Inferred reserves (growth)	.1	.11	5.4
Undiscovered resources	.6	.2	20.2
Total remaining petroleum	.900	.432	31.312

TABLE 7.108   Estimated Undiscovered Recoverable Conventional Oil
                Resources of the Atlantic Coast Region (in million barrels)

Area	Offshore (shelf 0–650 feet, slope below 650 feet)			Onshore
Florida peninsula				100–800 (300)[a]
North Atlantic	0–2,100	(400)	Shelf	
	0–3,800	(1,000)	Slope	
Mid-Atlantic	0–2,600	(800)	Shelf	
	0–7,600	(2,300)	Slope	
South Atlantic	0–300	(0)	Shelf	
Carolina trough	0–3,000	(600)	Slope	
Blake plateau	0–1,700	(300)	Slope	
Mean total		(5,400)		(300)

[a] See Table 7.2 for an explanation of estimates.
Source: See Table 7.2.

TABLE 7.109   Estimated Undiscovered Recoverable Conventional Natural Gas
                Resources of the Atlantic Coast Region (in trillion cubic feet)

Area	Offshore (shelf 0–650 feet, slope below 650 feet)			Onshore
Atlantic coastal plain				0–0.3 (0.1)[a]
Florida peninsula				0–0.1 (0)
North Atlantic	0–9.7	(2.5)	Shelf	
	0–12.5	(3.2)	Slope	
Mid-Atlantic	0–14.8	(5.6)	Shelf	
	0–26.9	(8.6)	Slope	
South Atlantic	0–1.0	(0.1)	Shelf	
Carolina trough	0–12.9	(2.8)	Slope	
Blake plateau	0–4.1	(0.7)	Slope	
Mean total		(23.5)		(0.1)

[a] See Table 7.2 for an explanation of estimates.
Source: See Table 7.2.

TABLE 7.110   United States Oil Production History (in million barrels)

Year	Production [a]	Reserves [a]	Reserves/Production Ratio
1969	3,188	29,632	9:1
1970	3,328	39,001	12:1
1971	3,298	38,063	12:1
1972	3,274	36,339	11:1
1973	3,179	35,300	11:1
1974	3,034	34,250	11:1
1975	2,901	32,682	11:1
1976	2,837	30,942	11:1
1977	2,865	29,486	10:1
1978	3,033	27,804	9:1
1979	2,958	27,051	9:1
1980	2,975	27,046	9:1
Total	36,870		

Reserve additions for the decade totaled 31,096 million barrels.

[a] Production and reserve totals include very small-volume producer states that were not included in regional data; therefore, regional totals will not add up to the exact final totals.

Source: See Table 7.1.

TABLE 7.111   Estimated Ultimate United States Oil Recovery
(in billion barrels)

Cumulative production (to end of 1980)	123.675
Proved reserves	27.046
Inferred reserves (growth)	26.3
Undiscovered resources	82.6
Total United States oil	259.621

TABLE 7.112   United States Gas Production History (in trillion cubic feet)

Year	Production[a]	Reserves[a]	Reserves/Production Ratio
1969	20.723	271.507	13:1
1970	21.961	286.743	13:1
1971	22.077	274.492	12:1
1972	22.512	261.614	12:1
1973	22.605	245.834	11:1
1974	21.318	233.194	11:1
1975	19.719	223.960	11:1
1976	19.542	211.973	11:1
1977	19.447	204.378	11:1
1978	19.311	195.653	10:1
1979	19.910	190.029	10:1
1980	19.333	188.161	10:1
Total	248.458		

Reserve additions for the decade totaled 144.389 trillion cubic feet.

[a] Production and reserve totals include very small-volume producer states that were not included in regional data; therefore, regional totals will not add up to the exact final totals.

Source: See Table 7.1.

TABLE 7.113   Estimated Ultimate United States Natural Gas Recovery (in trillion cubic feet)

Cumulative production (to end of 1980)	597.333
Proved reserves	188.161
Inferred reserves (growth)	174.4
Undiscovered resources	593.8
Total United States gas	1,553.694

TABLE 7.114   Estimated Ultimate United States Natural Gas Liquids
Recovery (in billion barrels)

Cumulative production (to end of 1980)	22.579
Proved reserves	6.728
Inferred reserves (growth)	6.59
Undiscovered resources	17.7
Total United States gas liquids	53.597

TABLE 7.115   Estimated Remaining United States Petroleum

	Oil (billion barrels)	Gas Liquids (billion barrels)	Natural Gas (trillion cubic feet)
Proved reserves	27.046	6.728	188.161
Inferred reserves (growth)	26.3	6.59	174.4
Undiscovered resources	82.6	17.7	593.8
Total remaining United States petroleum	135.946	31.018	956.361

TABLE 7.116   United States Oil Activity in 1980

Province Basin Type	Remaining Oil (million barrels)		Wells Drilled (1980)		Ratio of Wells/ Million Barrels
Alaska	13,800	disc.*	114	devel.*	1:121
Open downwarp	19,200	undisc.*	25	expl.*	1:768
Subduction					
Median	———		——		
Pull-apart	33,000	total	139	total	1:237
Rocky Mountains &	6,200	disc.	1,968	devel.	1:3
northern Great Plains	23,500	undisc.	1,114	expl.	1:21
Intracontinental composite	———		——		
Interior	29,700	total	3,082	total	1:10
Gulf Coast	11,700	disc.	5,158	devel.	1:2
Open downwarp	13,900	undisc.	1,045	expl.	1:13
Delta	25,600	total	6,203	total	1:4
West Texas &	10,700	disc.	8,298	devel.	1:1
eastern New Mexico	5,400	undisc.	1,456	expl.	1:4
Intracontinental composite	16,100	total	9,754	total	1:2
Pacific Coast	6,600	disc.	1,796	devel.	1:4
Subduction	8,200	undisc.	273	expl.	1:30
	14,800	total	2,069	total	1:7
Midcontinent	2,900	disc.	9,614	devel.	1:0.3
Intracontinental composite	4,200	undisc.	1,031	expl.	1:4
	7,100	total	10,645	total	1:0.7
Atlantic Coast	Negl.	disc.			
Pull-apart	5,700	undisc.	2	expl.	1:2,850
Michigan basin	1,000	disc.	289	devel.	1:3
Interior	1,100	undisc.	211	expl.	1:5
	2,100	total	500	total	1:4
Eastern Interior	300	disc.	2,598	devel.	1:0.1
Interior	800	undisc.	1,083	expl.	1:0.7
	1,100	total	3,681	total	1:0.3
Appalachian	300	disc.	2,352	devel.	1:0.1
Intracontinental composite	600	undisc.	36	expl.	1:17
	900	total	2,388	total	1:0.4

*disc. = discovered; undisc. = undiscovered; devel. = development; expl. = exploration.

Note: Unsuccessful exploratory wells are allocated proportionally to oil and gas.

Source: American Petroleum Institute, Quarterly Review of Drilling Statistics for the United States, Fourth Quarter and Annual Summary 1980, vol. 14, no. 4 (February 1981).

TABLE 7.117   United States Natural Gas Activity in 1980

Province Basin Type	Remaining Gas (trillion cubic feet)		Wells Drilled (1980)		Ratio of Wells/ Trillion Cubic Feet
Gulf Coast	188.9	disc.*	4,338	devl.*	1:0.04
Open downwarp	197.7	undisc.*	2,274	expl.*	1:0.09
Delta	386.6	total	6,612	total	1:0.06
Rocky Mountains &	31.6	disc.	1,892	devel.	1:0.02
northern Great Plains	136.2	undisc.	899	expl.	1:0.15
Intracontinental composite	167.8	total	2,791	total	1:0.06
Interior					
Alaska	38.8	disc.	3	devel.	1:12.9
Open downwarp	100.7	undisc.	0	expl.	
Subduction	139.5	total	3	total	1:46.5
Median					
Pull-apart					
Midcontinent	45.6	disc.	3,668	devel.	1:0.01
Intracontinental composite	44.4	undisc.	638	expl.	1:0.07
	90.0	total	4,306	total	1:0.02
West Texas &	30.6	disc.	1,721	devel.	1:0.02
eastern New Mexico	42.7	undisc.	974	expl.	1:0.04
Intracontinental composite	73.3	total	2,695	total	1:0.03
Appalachian	11.1	disc.	4,868	devel.	1:0.002
Intracontinental composite	20.2	undisc.	468	expl.	1:0.04
	31.3	total	5,336	total	1:0.006
Pacific Coast	8.4	disc.	42	devel.	1:0.2
Subduction	20.1	undisc.	159	expl.	1:0.1
	28.5	total	201	total	1:0.1
Atlantic Coast	Negl.	disc.			
Pull-apart	23.6	undisc.	2	expl.	1:11.8
Michigan basin	2.6	disc.	55	devel.	1:0.05
Interior	5.1	undisc.	110	expl.	1:0.05
	7.7	total	165	total	1:0.05
Eastern Interior	0.1	disc.	156	devel.	1:0.001
Interior	1.3	undisc.	140	expl.	1:0.009
	1.4	total	296	total	1:0.005

*disc. = discovered; undisc. = undiscovered; devel. = development; expl. = exploration.

Note: Unsuccessful exploratory wells are allocated proportionally to oil and gas.

Source: American Petroleum Institute, Quarterly Review of Drilling Statistics for the United States, Fourth Quarter and Annual Summary 1980, vol. 14, no. 4 (February 1981).

Notes

1. W. P. Borsge and I. L. Tailleur, "Northern Alaskan Petroleum Province," in *Future Petroleum Provinces of the United States—Their Geology and Potential,* Memoir 15 (Tulsa, Okla.: American Association of Petroleum Geologists, 1971), p. 94.

2. Ernst Mueller, "Alaskan Oil—The Energy Crisis and the Environment," *Arctic Bulletin,* vol. 1, no. 5 (1975), p. 186.

3. Howard M. Wilson, "Interior Seeks Industry Interest in Alaska Tracts," *Oil and Gas Journal* (May 18, 1981), pp. 48–49.

4. Howard M. Wilson, "Wildcatters Poised For the Beaufort Sea," *Oil and Gas Journal* (June 2, 1975), p. 100.

5. "New Alaskan Oil Will Add to Surplus on U.S. West Coast," *Petroleum Intelligence Weekly* (November 30, 1981).

6. J. Cordell Moore, "National Arctic Wildlife Range Seen as Best Potential For Huge Discovery," *Oil and Gas Journal* (August 6, 1979), p. 155.

7. Joseph P. Riva and James E. Mielke, *Polar Energy Resources Potential* (Report prepared for the Subcommittee on Energy Research, Development, and Demonstration and the Subcommittee on Energy Research,, Development, and Demonstration (Fossil Fuels) of the House Committee on Science and Technology [Committee Print 76-187] U.S. Congress, Washington, D.C., September 1976), p. 73.

8. A graben is an elongate, depressed crustal unit, bounded by faults on its long side. A horst is an elongate, uplifted crustal unit, bounded by faults on its long side.

9. Peter T. Hanley and William W. Wade, "Bering Basins Show Good Potential," *Offshore* (April 1981), p. 121.

10. *Ibid.,* p. 127.

11. *Ibid.,* p. 121–124.

12. *Ibid.,* p. 127.

13. Howard Wilson, "ARCO Dominates Gulf of Alaska Bidding," *Oil and Gas Journal* (October 27, 1980), p. 45.

14. George Gryc, "Summary of Potential Petroleum Resources of Region 1 [Alaska and Hawaii]—Alaska," in *Future Petroleum Provinces of the United States—Their Geology and Potential,* Memoir 15 (Tulsa, Okla.: American Association of Petroleum Geologists, 1971), p. 62.

15. All of the oil and gas production history tables in the chapter were derived from the following sources: American Petroleum Institute, American Gas Association, and Canadian Petroleum Association, *Reserves of Crude Oil, Natural Gas Liquids, and Natural Gas in the United States and Canada as of December 31, 1979,* vol. 34 (Washington, D.C., 1980); and Energy Information Administration, Office of Oil and Gas, *Principal Findings of the U.S. Crude Oil, Natural Gas, and Natural Gas Liquids Reserves,* 1980 Annual Report (Washington, D.C.: Department of Energy, 1981). The reserve estimates for

1980 were derived from combined American Petroleum Institute, American Gas Association, and Energy Information Administration figures.

16. United States inferred petroleum reserves were calculated by the U.S. Geological Survey at the end of 1979. (G. L. Dolton et al., *Estimates of Undiscovered Recoverable Conventional Resources of Oil and Gas in the United States,* Circular 860 [Reston, Va.: U.S. Geological Survey, 1981], pp. 22–25.) The inferred reserves in this chapter are extrapolated from these estimates using 1980 reserve values derived from combined Energy Information Administration, American Petroleum Institute, and American Gas Association estimates.

17. *Ibid.,* pp. 22–25 and 76–79. All of the undiscovered recoverable petroleum resource estimates (oil, gas, and natural gas liquids) in this chapter have been modified from this source.

18. "Executive Summary: Draft Report of the National Petroleum Council's Committee on Arctic Oil and Gas Resources," *U.S. Arctic Oil and Gas* (November 16, 1981), pp. 7–10.

19. Potential Gas Committee and Potential Gas Agency, *Potential Supply of Natural Gas in the United States (as of December 31, 1980)* (Golden, Colo.: Colorado School of Mines, 1981), pp. 31–32.

20. "Executive Summary," p. 9.

21. American Petroleum Institute, American Gas Association, and Canadian Petroleum Association, *Reserves of Crude Oil, Natural Gas Liquids, and Natural Gas,* p. 125.

22. B. C. Jones and D. W. Sears, "World Energy Developments, 1980: Alaska," *American Association of Petroleum Geologists Bulletin* (October 1981), pp. 1757–1765.

23. Dana B. Braislin, Douglas D. Hastings, and Parke D. Snavely, Jr., "Petroleum Potential of Western Oregon and Washington and Adjacent Continental Margin," in *Future Petroleum Provinces of the United States—Their Geology and Potential,* Memoir 15 (Tulsa, Okla.: American Association of Petroleum Geologists, 1971), pp. 229–230.

24. T. W. Dignes and D. Woltz, "World Energy Developments, 1980: West Coast," *American Association of Petroleum Geologists Bulletin* (October 1981), p. 1781.

25. Modified after Dolton et al., *Estimates of Undiscovered Recoverable Conventional Resources,* pp. 8–11.

26. E. N. Tiratsoo, *Oilfields of the World,* 2nd ed. (Houston, Texas: Gulf Publishing Company, 1976), p. 269.

27. *Ibid.,* pp. 273–275.

28. Robert R. Morrison et al., "Potential of Sacramento Valley Gas Province, California," in *Future Petroleum Provinces of the United States—Their Geology and Potential,* Memoir 15 (Tulsa, Okla.: American Association of Petroleum Geologists, 1971), pp. 329–338.

29. U.S., Department of the Interior, Bureau of Land Management, *Final Environmental Impact Statement: Proposed 1975 Outer Continental Shelf Oil and Gas General Lease Sale Offshore Southern California,* vol. 1, August 1975 (Washington, D.C.: Government Printing Office), pp. 68–71.

30. Dignes and Woltz, "West Coast," p. 1784.

31. "Work Picks Up Off California," *Offshore* (June 20, 1981), pp. 157–160.

32. Energy Information Administration, Office of Oil and Gas, *Principal Findings,* p. 4.

33. Ambrose L. Lyth, Jr., "Summary of Possible Future Petroleum Potential, Region 3, Western Rocky Mountains," in *Future Petroleum Provinces of the United States—Their Geology and Potential,* Memoir 15 (Tulsa, Okla.: American Association of Petroleum Geologists, 1971), p. 410.

34. John C. McCaslin, "Interest High Again in Utah's Great Salt Lake," *Oil and Gas Journal* (April 20, 1981), pp. 181–182.

35. John C. McCaslin, "Big Exploratory Task Ahead in Great Basin," *Oil and Gas Journal* (May 5, 1980), p. 303.

36. Philip F. Anschutz, "The Overthrust Belt: Will It Double U.S. Gas Reserves?" *World Oil* (January 1980), pp. 112–113.

37. *Ibid.,* p. 116.

38. Bob Tippee, "East Anschutz Ranch Shaping Up As Big Field," *Oil and Gas Journal* (November 2, 1981), p. 87.

39. R. D. TeSelle, D. D. Miller, C. B. Thames, Jr., and B. B. Sonderby, "Northern Rockies," *American Association of Petroleum Geologists Bulletin* (October 1981), p. 1793.

40. *Ibid.*

41. David R. Hoffman, "Williston—Reviving of a Giant Oil Province," *Oil and Gas Journal* (February 2, 1981), pp. 138–139.

42. L. A. McPeek, "Eastern Green River Basin: A Developing Giant Gas Supply From Deep Overpressured Upper Cretaceous Sandstones," *American Association of Petroleum Geologists Bulletin* (June 1981), pp. 1078–1098.

43. J. A. Momper and J. A. Williams, "Geochemical Exploration in the Powder River Basin," *Oil and Gas Journal* (December 10, 1979), pp. 129–134.

44. G. M. Stevenson and D. L. Baars, "Four Corners–Intermountain Area," *American Association of Petroleum Geologists Bulletin* (October 1981), pp. 1840–1841.

45. John Dolson, "Southern Rockies: Eastern and Northwestern Colorado," *American Association of Petroleum Geologists Bulletin* (October 1981), pp. 1836–1837.

46. Potential Gas Committee and Potential Gas Agency, *Potential Supply of Natural Gas,* pp. 28–30.

47. J. K. Hartman and Lee R. Woodward, "Future Petroleum Resources in Post-Mississippian Strata of North, Central, and West Texas and Eastern New Mexico," in *Future Petroleum Provinces of the United States—Their Geology and Potential,* Memoir 15 (Tulsa, Okla.: American Association of Petroleum Geologists, 1971), p. 752.

48. Tiratsoo, *Oilfields of the World,* p. 256.

49. Howard M. Wilson and Randy Sumpter, "Big Fields Help Slow Reserves Decline," *Oil and Gas Journal* (June 30, 1980), p. 36.

50. R. A. Diemer, W. R. Gibson, H. A. Miller, Jr., and L. D. Robbins, "West Texas and Eastern New Mexico," *American Association of Petroleum Geologists Bulletin* (October 1981), pp. 1843–1844.

51. David H. Root and Lawrence J. Drew, "The Patterns of Petroleum Discovery Rates," *American Scientist* (November-December 1979), pp. 648–652.

52. Kenneth O. McDowell and John S. Rives, II, "Louisiana Gulf Coast Onshore," *American Association of Petroleum Geologists Bulletin* (October 1981), pp. 1875–1877.

53. Laura J. Pankonien, "Operators Scramble to Tap Deep Gas in South Louisiana," *World Oil* (September 1979), pp. 55–62.

54. John C. McCaslin, "Exciting Tuscaloosa Trend Continues to Pay Off," *Oil and Gas Journal* (December 8, 1980), p. 127.

55. Potential Gas Committee and Potential Gas Agency, *Potential Supply of Natural Gas,* p. 27.

56. Deborah Cambre, A. G. Murphy, and Eugene Core, "Arkansas, Northern Louisiana, and East Texas," *American Association of Petroleum Geologists Bulletin* (October 1981), pp. 1862–1864.

57. P. D. Cate, "Southeastern States," *American Association of Petroleum Geologists Bulletin* (October 1981), pp. 1891–1893.

58. John C. McCaslin, "Anadarko Play Heads Busy Midcontinent Campaign," *Oil and Gas Journal* (May 25, 1981), pp. 161–171.

59. Robert A. Northcutt, "Oklahoma and the Panhandle of Texas," *American Association of Petroleum Geologists Bulletin* (October 1981), pp. 1848–1850.

60. Bailey Rascoe, Jr., "Western Kansas and Western Nebraska," in *Future Petroleum Provinces of the United States—Their Geology and Potential,* Memoir 15 (Tulsa, Okla.: American Association of Petroleum Geologists, 1971), pp. 1046–1054.

61. Shirley E. Paul et al., "North Mid-Continent," *American Association of Petroleum Geologists Bulletin* (October 1981), p. 1828.

62. Potential Gas Committee and Potential Gas Agency, *Potential Supply of Natural Gas,* p. 30.

63. Reserve estimates are derived from American Gas Association and Energy Information Administration data.

64. D. C. Bond et al., "Possible Future Petroleum Potential of Region 9— Illinois Basin, Cincinnati Arch, and Northern Mississippi Embayment," in *Future Petroleum Provinces of the United States—Their Geology and Potential,* Memoir 15 (Tulsa, Okla.: American Association of Petroleum Geologists, 1971), pp. 1214–1215.

65. Potential Gas Committee and Potential Gas Agency, *Potential Supply of Natural Gas,* p. 26.

66. Robert D. Lindau, Jacob Van Den Berg, G. L. Carpenter, and Edmund Nosow, "East Central States," *American Association of Petroleum Geologists Bulletin* (October 1981), pp. 1814–1816.

67. J. Labo, J. Cousins, W. G. Werner, and P. H. Pan, "Exploring for Silurian-Niagaran Pinnacle Reefs in the Southern Michigan Basin," *Oil and Gas Journal* (June 22, 1981), pp. 93–98.

68. "Michigan Deep Gas Strike Kicks Off Lease Play," *Oil and Gas Journal* (May 4, 1981), pp. 106–107.

69. D. Michael Bricker, "Michigan," *American Association of Petroleum Geologists Bulletin* (October 1981), pp. 1804–1806.

70. D. G. Patchen et al., "Northeastern United States," *American Association of Petroleum Geologists Bulletin* (October 1981), pp. 1896–1903.

71. W. Lynn Kreidler, "Petroleum Potential of New York," in *Future Petroleum Provinces of the United States—Their Geology and Potential,* Memoir 15 (Tulsa, Okla.: American Association of Petroleum Geologists, 1971), p. 1258.

72. Patchen et al., "Northeastern United States," p. 1897.

73. Burt Solomon, "U.S.G.S. Finds Extensive Gas Potential in Southeastern States," *Energy Daily* (October 9, 1979), p. 3.

74. Patrick Crow, "Gas Strikes Spark Play in Eastern Overthrust," *Oil and Gas Journal* (April 27, 1981), pp. 109–113.

75. F. A. Pontigo, Jr., A. V. Applegate, and J.H. Rooke, "S. Florida's Sunniland Oil Potential," *Oil and Gas Journal* (July 30, 1979), pp. 226–232.

76. J. Spivak and O. B. Shelburne, "Future Hydrocarbon Potential of Atlantic Coastal Province," in *Future Petroleum Provinces of the United States—Their Geology and Potential,* Memoir 15 (Tulsa, Okla.: American Association of Petroleum Geologists, 1971), p. 1295.

77. U.S., Department of the Interior, Bureau of Land Management, *Draft Environmental Statement: Proposed Increase in Acreage to Be Offered for Oil and Gas Leasing on the Outer Continental Shelf,* vol. 1, October 18, 1974 (Washington, D.C.: Government Printing Office), pp. 173–176.

78. Robert E. Mattick, "U.S. Atlantic Continental Margin, 1976–1981," *Oil and Gas Journal* (November 9, 1981), pp. 357–368.

79. *Ibid.,* p. 364.

80. John S. Schlee and John A. Grow, "Buried Carbonate Shelf Edge Beneath the Atlantic Continental Slope," *Oil and Gas Journal* (February 25, 1981), p. 156.

81. Drilling statistics were derived from the American Petroleum Institute's *Quarterly Review of Drilling Statistics for the United States,* Fourth Quarter and Annual Summary 1980, vol. 14, no. 4 (Feburary 1981).

82. Dolton et al., *Estimates of Undiscovered Recoverable Conventional Resources,* Appendix E, p. 80.

83. R. R. Johnson, "Drilling Activity in North America During 1980," *American Association of Petroleum Geologists Bulletin* (October 1981), pp. 1729–1730, 1753.

8

The Petroleum Prospects of the Soviet Union

A familiar pattern in petroleum-producing regions of the world is that production rises to a peak and then declines. Common experience has been that such declines are irreversible. The Soviets appear poised on the peak. To stabilize oil production through the 1980s, more oil will have to be found than was discovered in the 1970s or exploitation rates in existing fields, which are now nearing maximum production, will have to be further increased. Most of the new discoveries will have to be in the West Siberia basin (see Figure 8.1). However, if this basin behaves like others of its type, the production question will be not one of stabilization, but rather one of the timing and magnitude of decline.

History

The Caucasus and the Caspian Sea

The early history of oil production in Russia is associated mainly with the Baku region of Azerbaijan. In the thirteenth century, Marco Polo noted the export of naphtha by camel from the region. By 1825 Baku had about 125 wells producing about 25,400 barrels of oil annually.[1]

The production of Russian oil on a significant scale began in 1870, when Azerbaijan drilling at a depth of about 135 feet resulted in the discovery of a substantial field. Yearly oil production increased rapidly. In 1901 Russian output reached 84.5 million barrels, 81 million of which was from Azerbaijan. In the Caucasus, geological conditions were also favorable for the rapid increase of production, as relatively large fields were found close to the surface. Before World War II, about 80 percent of Soviet oil output was derived from the Caucasus region.

283

284

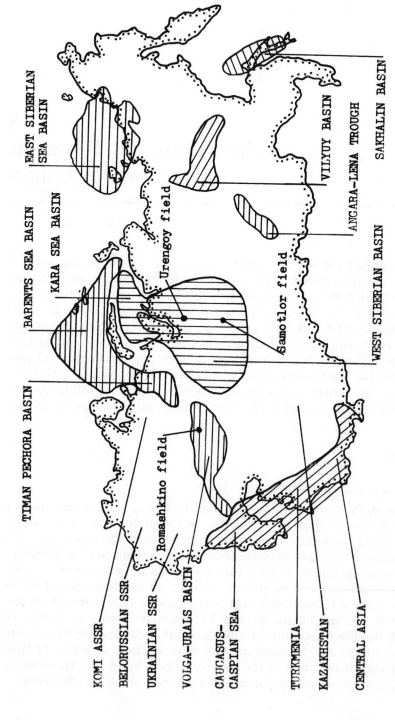

Figure 8.1. Map of sedimentary basins of the Soviet Union.

The production expansion, however, was not constant. In 1905 oil production declined to less than 59 million barrels and by 1913 the production achieved in 1901 had still not been surpassed.[2] The development of the oil industry in the Caucasus was retarded by strikes, World War I, and the Revolution and civil war. In 1917 the Bolshevik coup was strongly supported by the oil workers and a Soviet regime was initiated in Baku in November. Oil was nationalized in May 1918. During this period oil production dropped, mainly because of the loss of markets and also because of the fighting among the various factions and nationalities of the region. There had been considerable foreign investment (mostly from Britain and the United States) in the Baku oil industry. In 1913 foreign firms numbered 182, and about 60 percent of the industry was under the control of foreign investors.[3] The occupation of Azerbaijan by the Red Army in April 1920 brought to a final end the era of foreign participation in the Russian oil industry and placed the future of Caucasus oil in Soviet hands.

The new Soviet government attempted to increase oil production, but in spite of all efforts, output continued to fall. By 1928, however, a reversal was finally achieved and production reached 85 million barrels, near prerevolutionary levels. But Azerbaijan and the smaller fields in the North Caucasus and around Groznyy could not maintain the expansion, and exploration was initiated in other areas. Even so, in 1940, when national oil production had reached 229 million barrels, 163 million of this total still came from the Baku region.[4]

Baku subsequently declined and has never recovered this pre-World War II peak. Groznyy, discovered in 1890, peaked at 159 million barrels in 1972 and then declined rapidly to about 51 million barrels in 1977. The late peak was due to the discovery, in the later half of the 1950s, of new reserves in deeper Cretaceous horizons below the previously exploited Miocene. The Soviets projected continuously high Cretaceous production for Groznyy based upon the successes of the 1960s and set, in the Ninth Five Year Plan, a goal of 184 million barrels for 1975. Actual 1975 Groznyy production was about 66 million barrels.[5]

Although the Baku and Groznyy fields were not occupied during World War II, wartime exploration lagged and production declined. In an effort to restore production (Baku had dropped to less than one-half of its prewar magnitude) the Soviets began exploration and development in the Caspian Sea. The effort was successful. The Caspian Sea is the world's largest interior body of water. It is relatively shallow, with over 60 percent of its area 650 feet or less in depth. The most favorable portion of the Caspian Sea for petroleum is the Apsheron-Balkhan zone. This zone is made up of a chain of folded structures and includes a number of large offshore fields with middle Pliocene reservoirs. The first giant offshore Caspian Sea field was put into production in 1951. The field, estimated to have contained a total of 1.05 billion barrels of recoverable oil, is located at Neftyanyye Kamni, a reef area about twenty-five miles east of the Apsheron Peninsula of Baku. An entire town has been built over this area on more than 100 interconnecting platforms, linked with the mainland and to drilling rigs by a system of causeways, some of them 150 miles long. The Neftyanyye Kamni field, which has over 1,000 wells, is by far the most prolific single field yet found in the Caspian Sea.[6]

Neftyanyye Kamni production peaked at 51 million barrels of oil in 1970 and has since declined to less than half that amount. Another offshore field, Duvannyy-Bulla, replaced Neftyanyye Kamni as the leading Caspian oil producer in the early 1970s. Offshore development in the Baku region was not able to compensate for declining onshore production, however, and total Azerbaijan output (and production goals) continued to fall.

The decline continued until the development of the Muradkhanly field in 1977, which temporarily stabilized Azerbaijan production. However, the Baku region, which had been the world's dominant oil-producing region at the turn of the century, currently provides less than 5 percent of Soviet oil output and will not again play a major producing role.[7]

The Volga-Urals

The Baku-dominated phase of the Soviet oil industry came to an end in the mid-1950s with the development of the Volga-Urals intracontinental composite basin located between the Volga River and the Ural Mountains. The first oil discovery in this basin was made accidently in 1929 during potash exploration. Commercial oil production began in 1932 at Ishimbay. The early discoveries were in relatively shallow Permian and Upper Carboniferous rocks and were of limited production capacity. The most productive reservoirs were later found to be in Lower Carboniferous and Devonian rocks.[8] Dominant among the fields discovered in the Volga-Urals region was Romashkino, a super-giant that contained an estimated 14 billion barrels of recoverable oil.[9]

With the discovery of Romashkino in 1948 at Al'met'yevsk in Tatar ASSR (Autonomous Soviet Socialist Republic), the center of Soviet oil production began shifting to the Volga-Urals. In 1955, the giant Arlan field (with 2.8 billion barrels recoverable oil) was discovered in northwest Bashkiria and the Mukhanovo field (1.6 billion barrels of recoverable oil) in Kuybyshev Oblast. In 1958, Kuleshovka (760 million barrels of recoverable oil) was discovered, also in Kuybyshev Oblast.

By 1949 the expanding Volga-Urals oil production had compensated for the decline of the old Caucasus fields. In the mid-1950s the three leading producing regions of the Volga-Urals (Tatar ASSR, Bashkir ASSR, and Kuybyshev Oblast) had each surpassed the output level of Baku and had become, respectively, the first, second, and third leading oil-producing areas of the Soviet Union.[10]

By 1961, increased Volga-Urals oil production had raised the Soviet Union to the second-ranking oil-producing nation in the world (behind only the United States). In the mid-1960s the region accounted for over 70 percent of Soviet production. The beginning of oil operations in West Siberia reduced this percentage after 1965, but production did not peak in the Volga-Urals until 1975, when the Tatar ASSR reached its record production level of 764 million barrels of oil. Of this amount, the super-giant Romashkino produced 580 million barrels, or 76 percent, illustrating the critical importance of super-giant oil fields to major producing regions. Since the mid-1970s, production from the Volga-Urals province has been in decline.[11]

Also in the 1970s, oil output peaked in a number of other lesser Soviet oil-producing provinces. These include the Belorussian and Ukrainian republics in European Russia and Kazakhstan and Turk-menia.[12]

Other areas in which oil production is not declining are the Komi Republic, Sakhalin, Central Asia, and the Georgian Soviet Socialist Republic.[13] However, total current production from all these areas is not enough to offset the declines of the Volga-Urals region. The only area in which there is a major oil production increase is West Siberia.

Komi

One of the few older producing regions that appears to have even a modest prospect for short-term production increases is the Komi ASSR of northern European Russia. In this region, located in the Timan-Pechora intracontinental complex basin near the coast of the Barents Sea, the first commercial oil was produced in 1930. The early discoveries were mostly of heavy oil, but larger reserves of light oil were discovered in three fields in the 1960s. From 1960 to 1970 Komi oil production rose by a factor of 7 to 41 million barrels (plus an additional 15 million barrels of gas condensate).[14] The fields are located mainly between the Pechora River and its tributary, the Usa. Technical problems, including the early loss of reservoir pressure, which required water injection, delayed the start of full production until the mid-1970s. By 1976, oil production was 64 million barrels and by 1980 about 142 million barrels. However, the 1980 production was about 20 million barrels below the Five Year Plan goal for the region.

Despite the current growth in production, in the longer term Komi is expected to remain a minor producer. Beyond the major fields now in production, there appear to be limited prospects as exploration continues northward toward the coast of the Pechora Sea.[15]

The U.S. Geological Survey recently assessed the Timan-Pechora region, including the Barents Sea. Recoverable undiscovered oil was estimated to range from 8 to 58 billion barrels, with an expected value of 29 billion barrels. Undiscovered gas was projected to range from 179 to 498 trillion cubic feet, with an expected value of 325 trillion cubic feet. The geology of the region is not considered especially favorable for the accumulation of petroleum, but the area is huge, which in itself influences an assessment. Exploitation will be very difficult because of the Arctic conditions.[16]

West Siberia

European Russia (Caucasus, Caspian Sea, and Volga-Urals regions) dominated Soviet oil production into the 1960s. In the early 1960s, European Russia still accounted for 93 percent of Soviet oil production. The penetration of West Siberia began in 1960, with oil discoveries in the Shaim district of the Konda River valley. This location has become the farthest producing area to the west so far discovered in West Siberia, with a modest annual production of about 37 million barrels of oil.

The focus of exploration was shifted to the area around the Middle Ob' River Valley, 350 miles to the east, with the discovery in 1961 of the giant Ust'-Balyk field (which contained 2.3 billion barrels of recoverable oil) southwest of Surgut. The first oil from the Middle Ob' region was produced in 1964. Between 1961 and 1969 an intensive exploration effort in the region identified 59 fields including the super-giant Samotlor (in 1966), which contained some 14.5 billion barrels of recoverable oil. It is the largest field yet discovered in the Soviet Union (the tenth largest in the world) and is of the same order of magnitude as Romashkino, long the mainstay of Volga-Urals production. Other giant fields discovered, mainly in the area around Surgut, include West Surgut (730 million barrels), Pravdinsk (950 million barrels), and Mamontov (1.75 billion barrels).

Soviet plans called for the development of West Siberia's larger oil fields first, followed by medium-sized fields and smaller fields in that order. Within each category, priority was given to the fields with the highest-yield wells.[17] During the 1971 to 1975 Five Year Plan, the development of the West Siberia basin fields was dominated by the exploitation of the super-giant Samotlor. During that time this one field accounted for 66 percent of the production increase in West Siberia and 55 percent of the entire Soviet oil output growth.[18] In the second half of the 1970s, the development effort had to be shifted away from the Ob' valley and its large fields as more remote deposits in the swampy woodlands of the West Siberian plain were brought into production. The rate of development slowed as less accessible sites were exploited. This expansion proceeded northward from the Middle Ob' valley and also to the southeast in the Vasyugan Swamp of Tomsk Oblast.

West Siberia basin exploration drilling was somewhat curtailed in the late 1960s and early 1970s before again gaining. New West Siberian oil field discoveries fell from 15 in 1966 to 5 in 1974 and has since held at a rather low level. Goals for the discovery of new oil reserves fell short by considerable margins.[19] Development drilling, however, has increased steadily. Thus, the proportion of total Soviet oil drilling done in West Siberia and the annual footage reported for the region have both increased enormously.[20] In 1970, West Siberia accounted for only about 11 percent of the total footage drilled by the Ministry of Oil throughout the country. Total West Siberia drilling rose to about 36.5 percent of the nationwide total in 1978, and the 1979 goal was an additional increase to 44 percent. The 1980 plan called for over half (50.5 percent) of all Soviet drilling to be done in West Siberia.[21]

By far the largest drilling gains made in the West Siberia basin have been in development footage. By 1975, 31.2 percent of total development drilling was in West Siberia. The percentage increased to 43.6 in 1978 and was targeted at 55.8 percent in 1980. However, the 1979 West Siberia development drilling plan was only 82 percent fulfilled and the exploratory drilling only 70 percent fulfilled. Shortfalls were also expected in 1980.[22]

The 1979 total drilling goal for West Siberia was 24.6 million feet, which can be contrasted with 3.28 million feet in 1970. The 1979 total was divided into 21.04 million feet of development drilling and 3.56 million feet of exploration drilling. The actual achievement was 17.25 million feet of development drilling and 2.5 million feet of exploration drilling.[23]

Thus, although Soviet oil activities have concentrated more and more on the development of West Siberia fields, especially Samotlor, and exploration drilling is proportionally somewhat lower than in the United States, considerable exploration drilling has been accomplished. However, the amount of oil found per foot drilled (the discovery rate) has declined.

Current Production

Total Soviet production in 1980 was 4.390 billion barrels (12.03 million barrels per day).[24] This was an increase of 115 million barrels over 1979. These figures, however, are for petroleum liquids and include gas condensate production as well as oil. As the production of gas condensate has been increasing along with increasing natural gas production, the actual oil production increase is probably somewhat less than the 115 million barrels.

Of the 4.390 billion barrels total production, 2.278 billion barrels came from West Siberia (21.5 million barrels below the official West Siberia production goal), with Samotlor contributing 1.066 billion barrels (down from the 1.100 billion barrels produced in 1979).[25] Output from the Volga-Urals in 1980 was 1.404 billion barrels (down 67 million barrels from 1979), with a declining Romashkino contributing 500 million barrels to that total.

Thus, the West Siberia basin supplied 52 percent of total Soviet output in 1980, with Samotlor accounting for 47 percent of West Siberian oil. Samotlor produced 24 percent of total Soviet oil, and Romashkino added an additional 11 percent. Together, these two super-giant fields, now both in decline, accounted for 35 percent of 1980 Soviet oil production, which led the world.

In 1981 total Soviet petroleum liquids production was a reported 4.446 billion barrels (12.18 million barrels per day).[26] This represents an increase of only 56 million barrels over 1980, about 1 percent. This small increase might be accounted for by the increased production of gas condensate that has accompanied increased natural gas production, in which case the Soviet Union could have produced less oil in 1981 than in 1980. The West Siberia basin contribution to 1981 Soviet oil production rose to 55 percent.[27]

Summary

The Soviet Union has had a long history of oil production. A normal progression has occurred as several oil provinces have been discovered, become dominant producers, and then declined.[28] Petroleum development before World War II was gradual, as demand was low and technology limited. Exploration and production began in the Caucasus before 1900 and this rather modest province was the major domestic supplier until after World War II, accounting for 80 percent of Soviet production as late as 1940. The amount was moderate, the 229 million barrel total being only about one-fifth of the Samotlor field's current production. Caucasus production declined and after the war it became necessary to explore and develop the oil fields beneath the Caspian Sea. Some large fields were put into production, but they were not able to compensate for declining output onshore. Still, the Caucasus and the Caspian Sea provinces, when taken together, represent a considerable accumulation of oil, ranking eighth among world oil provinces with 23.8 billion barrels of known recovered and recoverable oil resources.[29]

The next oil province to be exploited was a giant. Known recovered and recoverable oil resources in the Volga-Urals province have been estimated at 40 billion barrels, making it the fourth-largest oil province in the world.[30] The super-giant field that has made this province so prolific is Romashkino, which contained 35 percent of the province's oil. This province, however, though it quickly raised the Soviet Union to the second-ranking world oil producer, peaked in twenty-five years and in fifteen years required assistance from West Siberia to maintain nationwide production increases. Currently, oil production from both Romashkino and the Volga-Urals province is in decline.

The West Siberia oil province is another giant, currently ranking third in the world with known recovered and recoverable oil resources estimated at 45 billion barrels.[31] The super-giant Samotlor field contained about 32 percent of this oil resource. The rapid exploitation of the third- and fourth-ranking oil provinces in the world (and specifically of the super-giant fields Romashkino and Samotlor), following the more gradual draining of the eighth-ranked province, propelled the Soviets in 1974 into the position of the world's leading oil producer. The 1971–1975 Five Year Plan concentrated on the successful development of Samotlor. The same plan, however, set a 1975 goal of 184 million barrels of oil from Groznyy (in the Caucasus), a projection that proved to be almost three times greater than the production actually achieved. In the second half of the 1970s an expanding West Siberia and a declining Volga-Urals province have combined to keep the Soviet Union the leading oil-producing nation in the world, with continually increasing production into the 1980s and a record level of 4.446 billion barrels of petroleum liquids in 1981.

Projections

The downward-revised Soviet goal for 1980 crude oil and condensate production of 4.453 billion barrels was not achieved. Likewise, the goal for 1981 of 4.490 billion barrels, indicating a leveling off in expectations, was not achieved. It is possible to put future Soviet production into perspective by using past production performance as well as the experience gained in other oil provinces throughout the world. The older Soviet provinces have followed a pattern familiar in petroleum-producing regions, namely, that production rises to a peak and then declines. Such declines have generally proved to be irreversible.

The Soviet situation can be best illustrated by using a projection of essentially level (12 million barrels per day or 4.38 billion barrels per year) oil and condensate production. As estimates regarding future Soviet petroleum liquids production range from sharply dropping to rising, a level production projection might be regarded as a rather conservative choice. It might be reasoned that, with technological advances and a very large area to explore, the Soviets surely will not do any worse in the future than they have done in the past. Even so, level production for the next five, ten, or twenty years is a very optimistic expectation. To illustrate this it is necessary to look back on the last decade during which the Soviets achieved a remarkable increase in oil production.

Several conclusions become apparent: 1. Soviet production increased from 2.555 to 4.390 billion barrels per year between 1970 and 1980. 2. Reserve information is generally not released by the Soviets, but it appears that during the same period there was little change in proved reserves from the 58 billion barrels estimated in 1970.[32] 3. If reserves have not increased, the reserves/production ratio (a measure of the rate of production) decreased from 23:1 to 13:1 and the Soviets are currently exploiting their oil reserves at almost twice the 1970 rate. 4. Total additions to reserves for the 1970s are estimated to be about 36.2 billion barrels (an average of 3.6 billion barrels per year). This can be compared to reserve additions of 31.1 billion barrels for the United States during the same period, which was characterized by a general production decline.   It can be seen that the increases in Soviet oil production during the 1970s were largely derived from the decrease in the reserves/production ratio. Although about as much oil was found as was produced during the past decade, the large increases in production were achieved by producing oil that for the most part was found during the twenty-five years preceding 1970.

A reserves/production ratio of 13:1 indicates a relatively mature petroleum industry. It is possible for the Soviet Union to further decrease its reserves/production ratio and thus realize additional production without commensurate reserve additions, but it would appear, given the state of available recovery technology and the remote location of the Siberian fields, that a production rate comparable to that of the United States (about 9:1) will not soon be achieved.

As the Soviets do not release oil reserve information on a regular basis, there has been much speculation about reserve size. Soviet reserve estimates are further complicated by differences in reserve definitions between customary Soviet and Western practice. The petroleum trade journals have indicated a considerable increase in Soviet reserves during 1980 and 1981, with reserves rising to 63 to 65 billion barrels. No giant field discoveries have been reported, and recent drilling has emphasized production rather than exploration, so if such increases are real and not technical corrections, they probably primarily represent upward revisions to reserves in existing fields due to development drilling rather than new discoveries. These higher reserve estimates would increase reserves/production ratios to 14:1 or 15:1. Other Soviet reserve estimates, however, are considerably lower. In 1977, the Central Intelligence Agency estimated that reserves were from 30 to 35 billion barrels, and the 1980 estimate of an independent geological consultant for the Soviet government put the Soviet oil reserve at about 43.5 billion barrels, resulting in a reserves/production ratio of about 10:1.[33]

The Soviet Union lacks advanced petroleum exploration and production technology. It sometimes takes as long as three to five years to drill a deep well that is often drilled in less than six months in the West.[34] Given inadequate drilling and production technology, a reserves/production ratio as low as 10:1 appears unlikely. The Soviet reserves/production ratio is probably between 13:1 and 15:1 and will vary within that range for the next several years. Thus, there is still some flexibility to increase production by producing existing reserves faster (reducing the reserves/production ratio), but not nearly as much as existed during the 1970s. A relatively stable reserves/production ratio seems more likely, at least through the 1980s.

To maintain production at about the current 12 million barrels per day (or 4.38 billion barrels per year) with a relatively stable reserves/production ratio will require, in general, that the Soviets add as much oil to reserves as they produce. This would mean that an average of about 4.38 billion barrels would have to be added to reserves each year to maintain level production until field development is intensive enough to reduce the resources/production ratio closer to a 10:1 annual withdrawal rate. As has been previously noted, an average of only 3.6 billion barrels of oil per year was added to reserves during the 1970s. Thus, an average of almost 20 percent more oil per year must be added than was added to reserves in the 1970s just to maintain level production, notwithstanding the fact that the past decade was characterized by a remarkable production increase. To stabilize production through the 1980s would require reserve additions of about 44 billion barrels of petroleum liquids, about 8 billion more than were added in the 1970s.

New discoveries will have to be giant fields to be produced fast enough to impact the next decade, given the long lead times in the vast frontier areas of the country. The primitive state of Soviet offshore technology all but precludes significant offshore Arctic production during the next decade, but new giant fields are more likely to be discovered in such remote undrilled areas.

An indication that the Soviets may expect reserve additions to decline, at least in the next five years, is contained in a statement by the deputy chairman of the state planning committee that average output from new West Siberia oil wells is expected to drop from the 679 barrels per day achieved during 1976–1980 to about 277 barrels per day during 1981–1985. This drop is ascribed to less favorable geological prospects in the new, more remote areas and development of smaller West Siberia fields. Approximately 60 new, relatively small fields are scheduled to be put into production during the next five years. These new fields have significantly poorer geological characteristics and lower per-well production potential than the Siberia fields developed in the 1970s.[35] If reserve additions decline and production can be maintained only by increasing output from existing fields, the production decline, when it finally occurs, will be even steeper than would have been the case otherwise because of the depleted reserves.

Prospects

Oil

It is clear that even to stabilize oil production new super-giant fields or even a new petroleum province is needed. The Soviet Union currently contains the eighth (Caucasus and Caspian), the fourth (Volga-Urals), and the third (West Siberia) largest petroleum provinces in the world. The country is vast, but in this century there appears to be no possibility of a new oil province overtaking West Siberia production in the manner that Azerbaijan was surpassed in the 1950s by the Volga-Urals, which was in turn surpassed by West Siberia in the 1970s.

Candidates for a new province would include East Siberia, a huge area in which exploration drilling continues. The Vilyuy intracontinental composite basin is a main prospect. Its deeper structures are complex and involve Precambrian and Paleozoic rocks. A Devonian salt dome province occurs in the western portion of the basin. Mesozoic rocks also occur. Gas has been found in Jurassic, Triassic, and Permian sandstones, but commercial quantities of oil have not yet been discovered.[36] East Siberia is not nearly as prospective as West Siberia.

The offshore area considered to have the greatest petroleum potential is the Kara Sea, located north of West Siberia. The Kara Sea basin is an extension of the West Siberia basin and is most prospective in rocks of Mesozoic age. Sea ice conditions, however, are extreme. The Soviet Union, relying heavily on foreign technology, has plans to begin drilling offshore in the Arctic within the next few years. Initial targets will probably be in the Pechora Sea (a large basin containing Paleozoic and Mesozoic sediments in which onshore commercial petroleum discoveries have been made). After gaining offshore experience in this area, the

TABLE 8.1   Original (Ultimate) Recoverable Oil Resources in the Volga-Urals Basin (in billion barrels)

Cumulative production	35
Proved reserves	5
Inferred reserves	5
Undiscovered resources (mean)	7 (range 3-14)
Total	52

Source: Charles D. Masters and James A. Peterson, "Assessment of Conventionally Recoverable Petroleum Resources, Volga-Urals Basin, U.S.S.R.," U.S. Geological Survey Open-File Report 81-1027 (Washington, D.C.: Department of the Interior, 1981), p. 7.

Soviets will probably move further north and east into the more hostile environment of the Kara Sea. Geophysical surveys in both seas have indicated significant petroleum potential, with gas predominating.[37]

The Volga-Urals intracontinental composite basin contains major petroleum reservoirs in sediments of Devonian and lower Carboniferous age. Upper Carboniferous and Permian sediments are also prospective and will contain many smaller fields, but giants are not expected. The province appears to be in a highly mature stage of exploration and significant new discoveries are not likely. Additional reserves are expected to be in small fields proximal to existing production and from the same part of the stratigraphic section.[38]

A recent geologically based assessment of Volga-Urals potential by the World Energy Resources Program of the United States Geological Survey is shown in Table 8.1.[39] The table indicates that 87 percent of the oil in the Volga-Urals basin is estimated to already have been discovered. Undiscovered Romashkino-sized fields are not considered likely to exist in the basin.

Thus, the most likely prospects for the discovery and exploitation of large oil fields, at least in this century, remain in West Siberia, and any discussion of future Soviet oil production potential should include undiscovered West Siberia oil resources. A recent geologically based assessment of West Siberian potential by the World Energy Resources Program of the United States Geological Survey is shown in Table 8.2.[40]

The West Siberia basin is an intracontinental complex basin, a type that tends to be not especially rich in petroleum but very gas prone. The West Siberia basin is prolific because of its enormous size rather than because of unusually favorable conditions for oil formation. A large portion of the petroleum is present in the form of biogenic gas

TABLE 8.2   Original (Ultimate) Recoverable Oil Resources in the West Siberia and Kara Sea Basins (in billion barrels)

Cumulative production	10
Proved reserves	35
Inferred reserves	30
Undiscovered resources (mean)	79 (range 25–176)
Total	154

Source: Charles D. Masters, "Assessment of Conventionally Recoverable Petroleum Resources of the West Siberian Basin and Kara Sea Basin, U.S.S.R., U.S. Geological Survey Open-File Report 81-1147 (Washington, D.C.: Department of the Interior, 1981), p. 6.

(predominantly methane) that is trapped beneath the permafrost in the northern part of the basin. Basins of this type, after an extended period of development, tend to have a closer balance between giant field and nongiant field reserves than do other types of basins. It is probable that most of the future discoveries will be of smaller fields, probably widely dispersed over the more central parts of the basin. However, some giant oil fields may still be discovered in the central and southern parts of the basin. Most of the oil in the West Siberia basin will be derived from Jurassic to Lower and middle Cretaceous sediments. Early Paleozoic rocks are also prospective but are not expected to be significant because of erratic distribution and limited source rock.[41]

Historical data indicates that giant fields in most basins are discovered rather early in the exploration cycle, as early drilling is targeted on the largest traps. This appears to have been the case in West Siberia. Although accessibility may have precluded some tests, it was probably not a major factor in drilling site selection. Exploration has been under way in West Siberia for twenty years and most of the large structures appear to have been drilled. The largest have often been found to contain gas, although 45 billion barrels of oil have been discovered and the growth of the discovered fields is expected to add an additional 30 billion barrels to the basin's recoverable reserve stock. About 79 billion barrels remain to be discovered. Because this half of the basin's total oil is more likely to be found in nongiant fields than in giants, and because nongiants take more time and resources to discover and exploit, it will be very difficult to maintain Soviet production with continued reserve additions from West Siberia. Still, this basin is the best hope, at least in this century, for new Soviet oil discoveries.

What is needed is a new super-giant field to augment Romashkino and Samotlor production, now in decline. Two possibilities have been suggested, Fedorovo and Salym, both located in the West Siberia basin.

Salym figured in the widely publicized December 1980 report by a consulting firm in Sweden that the Soviets had discovered a huge West Siberian oil field with reserves many times larger than those of the entire world. As absurd as this claim was, before it was refuted it shook investment circles and oil shares plunged on American stock exchanges.[42] What was referred to as the Bazhenov oil field in the report is a vast complex bituminous shale formation present over a large part of Siberia. Oil was discovered in the Bazhenov shale on the Salym uplift in 1967, and since that time about 80 wells have been drilled. As the structure is about 31 miles long and the wells were initially prolific, early estimates of reserves were in the super-giant range of Romashkino and Samotlor. However, the oil was found to be contained in the fractures of the shale and producing rates quickly dropped as the fracture systems were drained. Many field wells became dry and most others tested at only 1 to 50 barrels per day. Reserve estimates have been so drastically reduced that the Ministry of Geology recently stated that no significant reserves had been developed in the Salym area.[43] The Soviets have not as yet been able to determine how to economically extract the oil from the shale.

The Fedorovo field has become West Siberia's second-largest oil producer and the third-largest producer in the Soviet Union behind only Samotlor and Romashkino. It was discovered in 1971 and put into production in 1973. It was initially thought to be a giant field, but its maximum output was considered to be only about 51 million barrels per year. Fedorovo's 1975 production was slightly over 29 million barrels. Production was increased during the next few years and in 1979 an official Soviet report projected that Fedorovo's 1979 output would increase sharply to about 183 million barrels.[44] This amount was probably reached, and the 1980 production is considered to have been about 200 million barrels. Reserves in the field had been estimated at about 600 million barrels ranging to a possible 5 billion barrels. With over 600 wells drilled, current production indicates that 2 billion barrels is a more likely reserve volume. Thus, although Fedorovo is a most important producer, it does not appear to be the super-giant needed to stabilize Soviet production.

It is possible that giant oil fields exist and will be discovered in the West Siberia basin soon enough to be exploited to augment declining production elsewhere so that current Soviet oil production levels can be maintained for several more years. It is more likely, however, given the geological conditions and the mature state of petroleum development, that such discoveries will not be made (or will not be made in time), and a decline in Soviet oil production (which may already have begun in 1981) will occur.

TABLE 8.3   Original (Ultimate) Recoverable Natural Gas Resources in the
Volga-Urals Basin (in trillion cubic feet)

Cumulative production	data not available
Proved reserves	75
Inferred reserves	25
Undiscovered resources (mean)	63 (range 19–142)
Total	163+

Source: See Table 8.1.

Natural Gas

Unlike oil reserves, Soviet natural gas reserves are not classified as
secret. The Soviet Union contains the largest natural gas reserve in the
world, about 1,160 trillion cubic feet or about 6 times the reserve of
the United States. Gas production in the Volga-Urals is centered at
Orenburg, a super-giant gas field in the southeast part of the province.
The field, located in the more temperate European part of the Soviet
Union, was discovered in 1975 and contained reserves of an estimated
70 trillion cubic feet of gas. The gas is used for domestic industry and
for export to Eastern Europe through the 1,700-mile Orenburg pipeline.
A recent geologically based assessment of Volga-Urals natural gas is
shown in Table 8.3.[45]

The natural gas–prone region of the West Siberia basin extends for
over 1,000 miles from the Yamal Peninsula in the north to Tyumen
Oblast. Production in the area became substantial after 1970, when
pipelines were completed to transport the gas. The Tyumen fields are
of extreme importance to the Soviet gas industry. The increase in the
production of natural gas at Tyumen between 1975 and 1980 amounted
to more than 82 percent of the gas production increase for the entire
country. Gas production began in West Siberia in 1963 from several
small gas deposits located along the lower Ob' River. Three years later
the world's largest gas field was discovered at Urengoy east of Nadym
in the northern part of the basin. Proved Urengoy reserves are estimated
at 176.5 trillion cubic feet in the upper reservoirs with an additional
106 trillion cubic feet of inferred reserves in the lower horizons.[46] Other
super-giant West Siberian gas fields include Medvezhye (60 trillion
cubic feet of reserves), which began producing in 1972, and Yamburg
(71 trillion cubic feet of reserves), for which the controversial planned
pipeline to Western Europe was originally named. The gas for this
pipeline, however, at least initially will come from Urengoy.

A recent geologically based assessment of the natural gas situation

TABLE 8.4   Original (Ultimate) Recoverable Natural Gas Resources in the West Siberia and Kara Sea Basins (in trillion cubic feet)

Cumulative production	data not available
Proved reserves	400
Inferred reserves	260
Undiscovered resources (mean)	702 (range 235–1,519)
Total	1,362+

Source: See Table 8.2.

in the West Siberia basin and the Kara Sea basin is shown in Table 8.4.[47] The gas occurs mostly in the northern part of the basin because of a permafrost seal. It is biogenic in origin and is composed predominantly of methane.

Total Soviet gas production in 1981 was officially estimated at 16.38 trillion cubic feet, an increase of 7 percent over 1980.[48] Now second to the United States in gas production, but with a very much larger natural gas reserve, the Soviet Union is expected to pass the United States and become the world's leading gas producer in 1984. West Siberia accounts for over one-third of Soviet gas production. Future increases will depend upon further exploitation of this difficult area.

Conclusions

The oil sector is the most important fuel component within the Soviet energy system. The Soviet Union has rather recently become the world's leading oil producer. In 1980 about 3 million barrels of oil per day were exported both to allies and adversaries. Exports of oil to the West have accounted for over 50 percent of Soviet hard currency earnings since 1977. These export earnings are particularly important, as they allow the Soviets to purchase badly needed grain and advanced technology items. With so much riding on sustained oil production it is important to assess future oil production prospects. Even if natural gas exports expand as planned by the mid-1980s, they are not expected to bring in revenues equivalent to oil export earnings.[49]

During the long history of oil production in the Soviet Union, a pattern, familiar in all petroleum-producing regions, has recurred. As production rises to a peak and then irreversibly declines, oil provinces become dominant producers and then are surpassed. The Caucasus-Caspian was the first and the smallest. Discovered before 1900 and exploited rather slowly, it was not surpassed in production until the 1950s, when it was overtaken by Volga-Urals production. This oil province, the fourth largest in the world, quickly elevated the Soviet Union to the second-ranking world oil producer in 1961. However, Volga-Urals oil production peaked in twenty-five years and was surpassed in the 1970s by oil production from the West Siberia basin.

The rapid exploitation of the Volga-Urals province and the West Siberia province (the third-ranking world oil province), following the more gradual draining of the eighth-ranked province (the Caucasus-Caspian Sea), propelled the Soviet Union in 1974 into the position of the world's leading oil producer. Such rapid increases in production were largely the result of the exploitation of two super-giant fields. Together Samotlor and Romashkino account for more than one-third of Soviet oil production. Both fields are now in decline. In 1981, total Soviet petroleum liquids production was a reported 4.446 billion barrels, an increase of about 1 percent over the previous year. This small increase may be accounted for by the increase in gas condensate production, which accompanied increased natural gas production, and oil production may actually have fallen.

The Soviet oil production increase during the 1970s was apparently accomplished with little change in reserves. Thus, the Soviets are currently drawing down their oil reserves at almost double the 1970 rate. The increase in Soviet production was largely derived from the decrease in the reserves/production ratio. Although about as much oil was found as was produced during the past decade, the large increases in production were made possible by producing oil most of which was found during the twenty-five years that preceded 1970. It is possible to further decrease the reserves/production ratio and thus realize additional production without commensurate reserve additions, but it would appear technologically difficult for the Soviets to do so. Therefore, to stabilize oil production through the 1980s at essentially current reserves/production ratios would require reserve additions significantly higher than were achieved in the 1970s.

It is clear that even to stabilize production new super-giant fields or even a new petroleum province is needed. Candidates for a new province include East Siberia and the Kara Sea. Both, however, appear to be gas prone rather than oil prone. West Siberia most likely contains undiscovered oil reserves of sufficient volume to eventually elevate it to the second-ranking world oil province, behind only the Arabian-Iranian province. However, it is geologically a type of basin that tends to have a close balance of giant to nongiant reserves. Thus, it is probable that most of the future discoveries will be of smaller fields that are likely to be dispersed over the central and southern portions of the basin. This will result in slower and more costly exploitation. If West Siberia behaves like other basins of its type and the frontier provinces are found to contain mostly gas, only increased production rates in existing fields could maintain current production, and only for the short term. The question then would become not *will* Soviet oil production fall, but how soon and how far? Soviet oil production in the longer term will depend upon oil prospects in the frontier areas and the acquisition of technology adequate for their exploitation. It is doubtful, however, that another province as prolific as West Siberia will be found, and very large amounts of oil will be necessary to maintain the current high level of production. On the other hand, the prospects for increased natural gas production are quite good.

Notes

1. Iain F. Elliot, *The Soviet Union Energy Balance* (New York: Praeger Publishers, 1974), p. 70.

2. *Ibid.,* p. 71.

3. *Ibid.,* p. 70.

4. Leslie Dienes and Theodore Shabad, *The Soviet Energy System: Resource Use and Policies* (Washington, D.C.: V. H. Winston & Sons, and New York: John Wiley & Sons, 1979), p. 50.

5. *Ibid.,* p. 51.

6. Joseph P. Riva, Jr. *Soviet Offshore Oil and Gas: Soviet Oceans Development* (Report prepared for the Senate Committee on Commerce and Senate National Ocean Policy Study [Committee Print 69-315] U.S. Congress, Washington, D.C., October 1976), pp. 486–487.

7. Dienes and Shabad, *The Soviet Energy System.*

8. Charles D. Masters and James A. Peterson, "Assessment of Conventionally Recoverable Petroleum Resources, Volga-Urals Basin, U.S.S.R.," U.S. Geological Survey Open-File Report 81-1027 (Washington, D.C.: Department of the Interior, 1981), p. 7.

9. Richard Nehring, "The Outlook for World Oil Resources," *Oil and Gas Journal* (October 27, 1980), p. 172.

10. Dienes and Shabad, *The Soviet Energy System,* p. 52.

11. David H. Root and Lawrence J. Drew, "General Principles of the Petroleum Industry and Their Application to the U.S.S.R. Energy in Soviet Policy" (Report to the Subcommittee on International Trade, Finance, and Security Economics of the Joint Economic Committee [Committee Print 76-690] U.S. Congress, Washington, D.C., June 11, 1981), p. 136.

12. Dienes and Shabad, *The Soviet Energy System,* p. 53.

13. Arthur A. Meyerhoff, "Energy Base of the Communist-Socialist Countries," *American Scientist* (November-December 1981), p. 626.

14. Dienes and Shabad, *The Soviet Energy System,* p. 55.

15. *Ibid.,* p. 56.

16. Charles D. Masters, U.S.G.S. (personal communication).

17. Joseph P. Riva, Jr., and James E. Mielke, *Polar Energy Resources*

Potential (Report prepared for the Subcommittee on Energy Research, Development, and Demonstration and Subcommittee on Energy Research, Development, and Demonstration (Fossil Fuels), House Committee on Science and Technology [Committee Print 76-187], U.S. Congress, Washington, D.C., September 1976), p. 114.

18. Dienes and Shabad, *The Soviet Energy System,* p. 58.

19. "Soviet Geologists Rap Energy Planners," *Oil and Gas Journal* (February 5, 1979), p. 40.

20. "Soviets Push to Maintain Increases in Oil Output," *Oil and Gas Journal* (December 31, 1979), p. 38.

21. "Soviet Oil Decline Likely Despite Surge in Drilling," *Oil and Gas Journal* (September 8, 1980), p. 31.

22. *Ibid.*

23. *Ibid.*

24. "Communist Bloc Hikes Oil Flow Little More Than 2 Percent," *Oil and Gas Journal* (March 9, 1981), p. 47.

25. "Western Siberia Oil Output Falls Shy of Target," *Oil and Gas Journal* (March 16, 1981), p. 44.

26. "Soviets To Cut Oil Exports To Eastern Bloc," *Oil and Gas Journal* (January 25, 1982), p. 113.

27. "Oil Production Growth Rate Slowing Fast in Soviet Union," *Oil and Gas Journal* (November 30, 1981), p. 25.

28. Joseph P. Riva, Jr. *Soviet Petroleum Prospects: A Western Geologist's View: Energy in Soviet Policy* (Study prepared for the Subcommittee on International Trade, Finance, and Security Economics of the Joint Economic Committee, [Joint Committee Print 76-6900], U.S. Congress, Washington, D.C., June 1981), p. 122.

29. Nehring, "The Outlook for World Oil Resources."

30. Masters and Peterson, "Assessment of . . . Volga-Urals Basin," p. 6.

31. Charles D. Masters. "Assessment of Conventionally Recoverable Petroleum Resources of the West Siberian Basin and Kara Sea Basin, U.S.S.R.," U.S. Geological Survey Open-File Report 81-1147 (Washington, D.C.: Department of the Interior, 1981), p. 6.

32. "Communist Bloc Hikes Oil Flow"; and "Estimated World Proved Reserves of Crude Oil," *World Oil* (August 1971–August 1981; 11 issues).

33. Meyerhoff, "Energy Base," p. 628.

34. *Ibid.,* p. 626.

35. "Soviet Oil Decline Likely," p. 32.

36. Arthur A. Meyerhoff, "Soviet Arctic Oil and Gas: A Second Middle East," *Professional Engineer* (July 1975), pp. 30–31.

37. "Soviets Plan Arctic Work by 1982," *Offshore* (May 1980), p. 162.

38. Masters and Peterson, "Assessment of . . . Volga-Urals Basin," p. 7.

39. *Ibid.,* p. 6.

40. Masters, "Assessment of . . . West Siberian Basin and Kara Sea Basin."

41. *Ibid.,* p. 7.

42. Bill Paul, "Report on Soviet Oil Field Causes Stocks to Dive Before it is Disputed by Experts, *Wall Street Journal* (December 8, 1980).

43. "No Colossal New Oil Field in Siberia," *Oil and Gas Journal* (December 13, 1980), p. 42.

44. "Soviets Rank Fedorovskoye No. 2 in W. Siberia," *Oil and Gas Journal* (November 5, 1979), p. 44.

45. Masters and Peterson, "Assessment of . . . Volga-Urals Basin," p. 6.

46. Frank J. Gardner, "Russia Aiming to Surpass U.S. in Production of Gas," *Oil and Gas Journal* (April 14, 1975), p. 22.

47. Masters, "Assessment of . . . West Siberian Basin and Kara Sea Basin."

48. "Total Soviet Energy Output Logs Big Gain," *Oil and Gas Journal* (January 11, 1982), p. 135.

49. Jonathan B. Stein, "Soviet Energy," *Energy Policy* (December 1981), p. 302.

9

The Petroleum Prospects
of Mexico

Mexico's growing oil production is all that has kept its economy viable. Exports of petroleum products now earn over $14 billion annually, paying for food imports and servicing the large foreign debt. Framing an energy policy adequate to meet domestic needs and the crucial requirements of foreign exchange has proved a difficult task. There is an official oil production ceiling of 2.5 million barrels per day, with a margin of 10 percent that allows an increase to 2.75 million barrels per day. Although there is concern that production increases would further aggravate inflation, current Mexican oil reserves would support much higher production levels, and there is also excellent potential for additional significant discoveries.

History

The first oil well in Mexico was drilled in 1869 near the Furbero seeps in Vera Cruz (see Figure 9.1). Commercial oil production was not established until 1904, when the La Paz exploration well was drilled in San Luis Potosi. Subsequent oil development was very rapid, with many large accumulations discovered in the next few years. Most of the new fields were in southern Mexico, along a geological trend named the Golden Lane. In 1921, maximum production reached a peak of 193 million barrels, or about 530,000 barrels per day, which represented 25 percent of world production.[1] A sharp decline followed due to the rapid exhaustion of the Golden Lane reservoirs, which are composed of highly permeable limestones. In 1938, total annual output had fallen to only 39 million barrels of oil, or about 107,000 barrels per day.

In the early 1970s Mexico had to become an importer of oil to meet

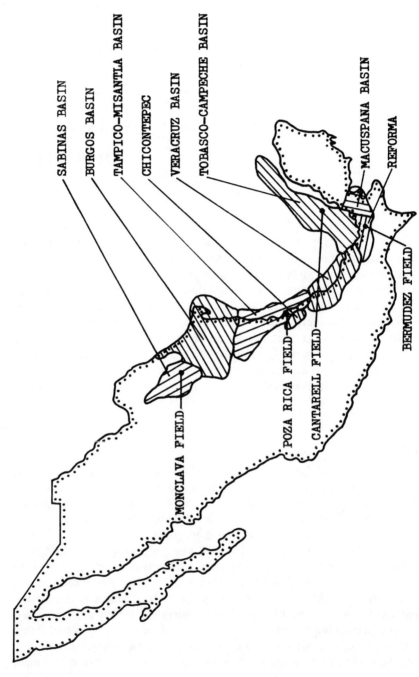

Figure 9.1. Map of sedimentary basins of Mexico.

TABLE 9.1   Original (Ultimate) Recoverable Petroleum Resources in the Burgos Basin, Mexico

	Oil *(billion barrels)*	*Gas* *(trillion cubic feet)*
Cumulative production	0.1	4.0
Proved reserves	0.3	5.0
Undiscovered resources (mean)	0.5 (range 0.1–1.2)	5.5 (range 1.6–12.5)
Total	0.9	14.5

Source: Charles D. Masters and James A. Peterson, "Assessment of Conventionally Recoverable Petroleum Resources of Northeastern Mexico," U.S. Geological Survey Open-File Report 81-1143 (Washington, D.C.: Department of the Interior, 1981), p. 7.

its domestic consumption requirements. However, the discovery of new oil reserves near Tampico and Tuxpan during 1968–1970 and in the Chiapas-Tabasco-Campeche area after 1972 made Mexico again self-sufficient in oil by 1974.[2] By the fall of 1974, Mexico resumed exporting, at the rate of 35,000 barrels of crude oil per day. Average daily exports increased to 250,000 barrels in early 1978, and Mexico now has export contracts for 1.4 million barrels of crude oil per day. Current production has surpassed 2.5 million barrels per day.[3]

Areas of Petroleum Accumulation

Northeastern Area

The Burgos basin in northeastern Mexico has accounted for substantial petroleum production. The area is geologically a part of the Rio Grande embayment of the Gulf Coast open extracontinental downwarp basin. The basin extends into south Texas. Structural traps consist mainly of gentle folds with northwest-southeast axes. The main petroleum-producing formations are Eocene and Oligocene sandstones. Giant fields are not likely in the Burgos basin because of a mature exploration situation and the deltaic environment during Tertiary deposition. Overall stratigraphic thickness of Tertiary rocks may exceed 40,000 feet.[4] Deeper gas was discovered in the basin in early 1979, and about half of the gas reserves in the basin have been found since 1975.

A recent geologically based assessment of the conventionally recoverable petroleum resources of the Burgos basin is shown in Table 9.1.[5] About 2.5 to 3.0 trillion cubic feet of the gas reserves in the basin are contained in the Reynosa field.

TABLE 9.2   Original (Ultimate) Recoverable Petroleum Resources in the
Sabinas and Parras Basins, Mexico

	Oil *(billion barrels)*	*Gas* *(trillion cubic feet)*
Cumulative production	0	data not available
Proved reserves	0	2
Undiscovered resources (mean)	0.8 (range 0–2.9)	55 (range 18–122)
Total	0.8	57+

Source: See Table 9.1.

Commercial gas fields have also been discovered outside of the Gulf coastal plain in the Sabinas basin of Coahuila and Nuevo Leon near the United States border. Petroleum exploration in the Sabinas basin is in an early stage. The discoveries so far have been of gas with high initial flow rates. No oil discoveries have yet been reported, but geological investigations suggest that oil accumulations are likely. Giant fields are likely to occur in the Sabinas and nearby Parras basins. The Sabinas basin contains over 65 structures, about half of which are almost 20 miles long. Gas discoveries have been made in at least 9 of these structures. One 22-mile-long field, the Monclava–Amuleto–Buena Suerte, has a reported 2 trillion cubic feet of proved gas reserves. The discoveries to date have been in long, narrow anticlines with production coming from Jurassic and Cretaceous dolomites. The stratigraphic sequence includes salt beds and coal seams. The source rocks are probably of Jurassic age. Overall thickness of Mesozoic rock units may be between 30,000 and 40,000 feet. The Sabinas and Parras basins contain geological sections similar to those in the western United States from the Overthrust Belt east to the craton.

A recent geologically based assessment of the conventionally recoverable petroleum resources of the Sabinas and Parras basins is shown in Table 9.2.[6] The Sabinas basin could possibly be one of the largest sources of gas in North America.

Central Area

The central area includes the Tampico-Misantla basin, the Veracruz basin, and Chicontepec and similar areas (see Figure 9.1).

The Tampico-Misantla basin is, in part, an extension of the Gulf Coast basin. The northern fields in this basin occur on a south-plunging prolongation of the Sierra Tamaulipas mountain anticline. The principal oil reservoirs are Cretaceous limestones. The limestones are dense and production depends largely on the existence of secondary porosity and permeability caused by fissuring, solution, or jointing. Thus, oil wells vary greatly in productivity even over short distances due to varying local reservoir conditions. The oil is heavy with a high sulfur content. Oil was first discovered in the area in 1904, and since that time a number of fields have been found. The region contains many oil seeps. Ebano, the first field discovered in the area, proved to be a giant.

The Tuxpan fields were discovered to the northwest and west of Tuxpan, along a narrow, arcuate, anticlinal ridge extending 51 miles from Dos Bocas in the northeast to San Isidro in the south. Asphaltic seeps are common in this region and are often associated with igneous intrusions. The arcuate anticlinal ridge is known as the Golden Lane because of the many important oil fields it contained. These fields occurred on structural highs separated from each other by faults or local saddles. The limestone formation that formed the principal oil reservoir was a barrier reef in Early Cretaceous time. At present it consists of about 8,000 feet of massively bedded limestone that outcrops at the surface in a nearby mountain range. Consequently, the oil fields have a strong natural water drive. In addition, the limestone was probably the source rock for the oil it contains. Primary porosity is principally due to hollow fossil casts and shell breccia. Fracturing, faulting, and solution cavities have also added greatly to the porosity and permeability of the reservoir rock.

Development of the Golden Lane fields began in 1910. One of the first fields discovered, Naranjos–Cerro Azul, was a giant field. Until 1933, these fields were the most important in Mexico, as almost every well drilled into one was a large oil producer. Some wells, in fact, produced tremendous flows of oil; the initial output of Cerro Azul Number 4 well was a reported 260,000 barrels of oil per day. Within this belt more than 1.35 billion barrels of oil were produced. Production was rapid and increased steadily to a peak in 1921.

The wells, however, tended to produce salt water suddenly when they were reached by the oil-water contact, and the great reservoir permeability that permitted rapid exploitation also led to an extraordinary rapid production decline. By 1919, salt water already had begun to invade the fields.[7]

The output from the Golden Lane fields has declined and other similar, but mostly smaller, fields have been discovered on roughly parallel limestone structures both to the west and, more recently, to the east in the Gulf of Mexico. The Golden Lane structure continues in an arc extending to the southeast between Tuxpan and Nautla. The giant Poza Rica field was discovered in 1928 on this structural trend on a locally plunging anticline. The reservoir is a porous limestone of

Early Cretaceous age. The field currently produces at a rate of about 34,250 barrels of oil per day. The second-largest field on the same structural trend is San Andres, currently producing about 22,500 barrels of oil per day.

Before the development of the Reforma area, the Tampico-Misantla basin provided the bulk of Mexico's oil production. This well-explored basin appears to have almost no potential for future major petroleum discoveries. Reserves are about 3.5 billion barrels of oil and 2.3 trillion cubic feet of gas.

The Veracruz basin lies to the south of the Tampico-Misantla basin and differs from it stratigraphically because of facies changes in the Cretaceous rocks. The basin is an extension of the Reforma-Compeche reef trend. The oil reservoirs are Cretaceous limestones and there is gas output from Oligocene sandstones. Oil was first discovered in 1953, but the fields were small. Hydrocarbon potential is limited in the Veracruz basin because reservoirs and traps are likely to be small. There is an overthrust region that may have petroleum potential, but it occurs at great depth. Proved reserves are estimated to be 200 million barrels of oil and 1.1 trillion cubic feet of gas.

Pemex, the Mexican national oil company, has reported plans to further develop the large Chicontepec region, located west of the Tampico–Poza Rica producing area on the coastal plain of the Gulf of Mexico. Geologically, Chicontepec consists of large erosional canyons that became receptacles for thick shale, carbonate, and sand deposits. The resulting Tertiary deposits contain well-sorted and tightly cemented sands in bars and channels. The reservoir rocks are Eocene sandstones occurring in three zones. Wells in these widespread sandstones are less prolific than those in other Mexican areas.

Chicontepec productive capacity is currently about 14,700 barrels of oil and 24.4 million cubic feet of gas per day.[8] The area is estimated to contain at least 100 billion barrels of in-place oil, and Pemex projects a possible ultimate recovery of about 10 billion barrels. However, poor reservoir characteristics, including low porosity and permeability and high water saturation, make a projection of a 10 percent recovery dubious. About 16,000 wells will be needed to fully develop the region. To date, 37 fields have been identified. Based on considerations of reservoir quality, the availability of all but a minor amount of Chicontepec petroleum oil resources must be discounted.

Southern Area

The Macuspana basin is located between the Reforma Mesozoic fields and the Yucatan platform and extends northward to join the Campeche marine platform. Three giant gas and gas condensate fields have been discovered in this basin to date (José Colomo, Chilapilla, and Hormiguero). Production is from anticlinal traps with Miocene sandstone reservoirs.[9]

About 23 giant and large oil and gas condensate fields, the most important petroleum deposits in Mexico, have been discovered since 1972 on the Reforma-Campeche shelf trend in the southern area. Fifteen of the fields are onshore and 8 are offshore. The reservoirs are mainly Late Jurassic and Early–mid-Cretaceous bank-edge talus, but offshore Paleocene production has also been found. The Reforma-Campeche trend is stratigraphically analogous to the equivalent-age Golden Lane trend farther to the north.

The Golden Lane and Reforma-Campeche trends are Late Jurassic to mid-Cretaceous giant reef atolls formed in an open downwarped basin and associated with three facies: a lagoonal and bank-edge shallow water carbonate and evaporite facies, a bank-edge carbonate talus or reef-debris facies, and a deepwater carbonate facies. Structurally, the area is very complex. The traps are horst and graben features that caused the localization of bank-edge debris (the reservoir rock) in specific areas. The reservoir rock is fractured in part by salt movement, which is apparently necessary for permeability. The microfracturing system, created by salt movement, produces huge reservoirs with thick effective oil columns. The seals for the fields are provided by shales of Cretaceous and Miocene age onshore. Paleocene-age seals are present offshore in fields in which the Paleocene bank-edge talus is productive.[10] The source of the petroleum appears to be Jurassic marine deposits and possibly also middle Cretaceous organic-rich shales and limestones. In the Reforma trend, onshore, oil predominates in the center of the bank-edge talus trend, and gas and condensate are found mostly along the flanks.

The Reforma region is located onshore adjacent to the Macuspana basin. The first giant field in the region, Sitio Grande, was discovered in 1972. The field is in an anticline about 6 miles long and 4 miles wide, an asymmetric structure cut by two systems of normal faults. Closure averages about 1,600 feet in the Cretaceous carbonate reservoir. The productive zone ranges in thickness from 325 to 500 feet. Sitio Grande, with the help of a waterflood, has produced over 200 million barrels of oil and 258 million cubic feet of gas. Oil is currently being produced at the rate of about 76,000 barrels per day. Oil reserves are estimated at about 600 million barrels.

Cactus, discovered in the same year, is located about eight miles north of Sitio Grande. It is also an anticline of similar size with two well-defined fault systems. Production is from the Cretaceous with a productive zone of about 3,250 feet. Cumulative production has exceeded

237 million barrels of oil and 1 trillion cubic feet of gas. Oil reserves are estimated at about 1.5 billion barrels, making Cactus a giant field. Current production is about 51,000 barrels of oil per day.

An extensive exploration campaign followed the Sitio Grande and Cactus discoveries in 1972. A number of discoveries were made in the mid-1970s that at first were thought to be separate fields and later found to be part of a single field, the Antonio J. Bermudez complex. The structure is 10 miles long by 8.5 miles wide and is cut by numerous normal faults. In the middle of the structure a reverse fault divides the complex, the northern block overriding the southern block. Major faults striking northwest-southeast dissect the structure, causing the productive horizons to have different structural and oil-water contact levels. In one section of the field there are about 4,000 feet of oil-saturated rocks.[11]

Cumulative production exceeds 2 billion barrels, with an expected recovery of an additional 5 billion barrels, making Bermudez a super-giant field ranking among the 37 largest fields in the world. Cumulative gas production is about 3 trillion cubic feet and current oil production about 500,000 barrels per day. A large waterflood is under way, which is expected to enhance recovery. When completed the program will include the injection of more than 1 million barrels of water per day.[12]

The giant Iris-Giraldas field, about 7 by 5 miles, was discovered in 1977. Petroleum is produced from Cretaceous dolomites. Cumulative production is about 21 million barrels of oil, with current production about 41,000 barrels per day. Reserves are estimated to be 1.5 billion barrels of oil.[13]

The Campeche offshore area has been explored by reflection seismology with a large number of anticlinal structures thus defined. Over 20 successful wildcats have been drilled into some of the structures. The discovery well was Chac 1, completed in lower Paleocene dolomitic breccias at about 16,500 feet in 1976. The well is located on the southeastern flank of a 19-by-6-mile anticline. The Akal field is located on the top of the anticline and the Nohoch field on the south flank. These fields are collectively known as the Cantarell complex, the most productive field in Mexico, with 44 wells flowing a total of over 1 million barrels per day. Cantarell is a super-giant field with cumulative production of 432 million barrels of oil and proved reserves estimated at 8 billion barrels. Cumulative gas production exceeds 22 billion cubic feet.[14] Inferred reserves for Cantarell have been projected at about 11.6 billion barrels. If 20 billion barrels of oil prove to be recoverable from Cantarell, it would rank in size in the top half-dozen fields in the world.

TABLE 9.3 Original (Ultimate) Recoverable Petroleum Resources in Southeastern Mexico, Northern Guatemala, and Belize

	Oil (billion barrels)	*Gas (trillion cubic feet)*
Cumulative production	6	8
Reserves	53	65
Undiscovered resources (mean)	78 (range 26–168)	120 (range 42–254)
Total	137	193

Source: Charles D. Masters and James A. Peterson, "Assessment of Conventionally Recoverable Resources of Southeastern Mexico, Northern Guatemala, and Belize," U.S. Geological Survey Open-File Report 81-1144 (Washington, D.C.: Department of the Interior, 1981), p. 7.

Waterfloods are planned for the offshore Campeche fields to assist in petroleum recovery.

North of Cantarell, petroleum prospects are limited by loss of fracturing enhanced by salt movement and loss of overlying cap rock. Also, there may be flushing of oil from this area from the emergent Yucatan structural high to Jurassic and Cretaceous outcrops on the seafloor. Areas that remain prospective include: structures immediately northwest and southwest of Cantarell, including a possible horst block high in the syncline between Cantarell and Reforma; additional Reforma structures; and stratigraphic prospects seaward and against the reef bank.[15] (Although over 30 structures have been reported as having been identified by initial seismic surveys by Pemex, oil occurrence and entrapment result from a combination of many geological factors, each having unique areal limits. Therefore, undiscovered resources cannot be estimated by counting structures and assuming a percentage will be commercial producers.) The discovery of light crude oil in the Abkatun field, southwest of Cantarell, has provided further indications of a potentially productive belt between Cantarell and Reforma and has caused Mexico to again raise its estimated petroleum reserves.

A geologically based assessment of the conventionally recoverable petroleum resources of southeastern Mexico, northern Guatemala, and Belize has been recently completed. The Guatemala area appears to be located over a back-reef (lagoonal) facies not likely to contain major petroleum resources. Thus, the following estimates (see Table 9.3) are almost entirely associated with southeastern Mexico, including the Veracruz basin.[16]

TABLE 9.4   Original (Ultimate) Recoverable Petroleum Resources in Mexico

	Oil *(billion barrels)*	*Gas* *(trillion cubic feet)*
Cumulative production	8.3	12+
Reserves	57.0	75.4
Undiscovered resources (mean)	79.3	180.5
Total	144.6	267.9+

Sources: See Tables 9.1, 9.2, and 9.3; also, *Oil and Gas Journal* (December 29, 1980, and December 28, 1981); and *World Oil* (August 1981).

Oil and Gas Production, Reserves, and Resources

Oil and gas production can best be viewed in the context of total Mexican petroleum reserves and resources (see Table 9.4). Mexican oil production in 1981 may have exceeded the official production ceiling of 2.75 million barrels per day. Pemex has estimated that the average production rate may have approached 2.79 million barrels per day because of increased production in the Bay of Campeche where output was estimated to have risen from 1.3 to 1.6 million barrels per day during the year. Reforma production was estimated to have averaged 0.93 million barrels per day during 1981, with the balance coming from other areas.[17] Mexico has contracts to export about 1.4 million barrels of oil per day with a target of 1.5 million barrels per day. As Mexican oil reserves are about 60 percent heavy oil and 40 percent lighter oils, a new policy of the government is to significantly increase the refining of the heavy-grade oils domestically to permit the increased export of light crudes, which are more in demand.

With production averaging 2.79 million barrels per day, Mexico would have produced 1.018 billion barrels of oil in 1981. With reserves estimated at 57 billion barrels, this represents a reserves/production ratio of 56:1. With such a large reserve, Mexico could considerably increase oil production. Given the technical ability of Pemex, a 15:1 reserves/production ratio could probably be achieved. This would translate to a production increase to 3.8 billion barrels per year or 10.4 million barrels per day. It is unlikely that Mexico will permit its oil production rate to reach this level, however, because official plans call for conservation of the petroleum resource and the diversification of Mexican industry into nonpetroleum areas.

Gas production in 1981 averaged 4.1 billion cubic feet per day (1.496

trillion cubic feet per year), of which 3.16 billion cubic feet per day was associated with oil production. The reserves/production ratio of Mexican gas is 50:1. Gas supplies will increase as turbocompressor modules on order for the Bay of Campeche allow the gathering of an additional 1.1 billion cubic feet of gas from Campeche fields. The gas is now being flared. It is possible that Mexico will sell additional gas to the United States because it may not easily be able to absorb this large new supply domestically.

Summary

Mexican oil reserves are by far the largest in the Western Hemisphere and rank fifth in the world. Mexican production, however, is relatively small, reflecting incomplete development of many of the recently discovered giant fields and a government policy to moderate production increases.

Oil was discovered early in Mexico, with the first commercial production in 1904. Development was rapid as large fields were discovered along a geological trend called the Golden Lane. By 1921 Mexico accounted for 25 percent of world oil production. Some of the wells, particularly those in Cerro Azul (a giant field), produced tremendous flows of oil. However, the great reservoir permeability that permitted such flows also led to extraordinary production declines and by 1919 salt water had begun to invade the fields.

The Golden Lane trend continues southeast, where the giant Poza Rica field was discovered in 1928 on a locally plunging anticline. Before the development of the Reforma area to the south, the Golden Lane region provided the bulk of Mexico's oil production, but output continued to fall and in the early 1970s Mexico had to import oil to meet its domestic needs.

The discovery of giant petroleum accumulations in the Reforma-Campeche region of southeastern Mexico after 1972 rapidly propelled Mexico back into the ranks of the oil exporters. In the fall of 1974 exports were resumed and now pay for both food imports and service on the large foreign debt.

The Reforma-Campeche shelf area has so far been found to contain 15 giant and large oil, gas, and condensate fields onshore and 8 offshore. Among these are two super-giants: Bermudez, onshore in Reforma, and Cantarell, offshore in Campeche. The Reforma-Campeche area accounts for about 87 percent of Mexico's total oil production and contains by far the best prospects for additional oil discoveries.

The Reforma-Campeche reservoirs are late Mesozoic and early Cenozoic giant reef atolls and banks formed in an open downwarped extracontinental basin. Oil and gas are produced from the bank-edge talus or reef debris facies. Production rates are high and it is possible that the decline experience of the Golden Lane fields (which occur in similar deposits) may be repeated in portions of certain Reforma-Campeche fields. The fields are quite deep (12,000 to 17,000 feet), which explains their late discovery and exploitation. It was necessary for Pemex to acquire the technology to explore and produce onshore and offshore fields at these depths before the region could be developed.

Mexican gas production is relatively small when compared to its sizable gas reserves. Steps are being taken to recover additional associated gas, which is being flared. This may lead to increased sale of gas to the United States, as Mexico may not be able to absorb this large new supply domestically. Excellent prospects for gas occur in the Sabinas basin in the northeastern part of Mexico. The basin could possibly be one of the largest sources of gas in North America.

Mexico possesses a very significant petroleum reserve and excellent prospects for large additional discoveries. The decision has not yet been made as to the pace and the extent of domestic oil and gas development. One political faction feels that foreign capital should be utilized to accelerate production and exports, but another feels that the development of domestic and agrarian enterprises should be granted a higher priority. The resolution of this debate, against a backdrop of spiraling population (growth rate of 3 percent per year) and unemployment and under-employment of 45–50 percent, will determine whether Mexico will become one of the richest and/or one of the most unstable of Third World oil exporters.

Notes

1. E. N. Tiratsoo, *Oilfields of the World,* 2nd ed. (Houston, Texas: Gulf Publishing Company, 1976), p. 290.

2. Joseph P. Riva, *The Geology of Mexico's Oil and Gas Resources: Mexico's Oil and Gas Policy: An Analysis* (Report prepared for the Senate Committee on Foreign Relations and the Joint Economic Committee [Joint Committee Print 36-950], U.S. Congress, Washington D.C., December 1978), pp. 59–67.

3. Walter Goldstein, "Political Regulation of Mexico's Oil Boom," *Energy Policy* (December 1981), p. 251.

4. Charles D. Masters and James A. Peterson, "Assessment of Conventionally Recoverable Petroleum Resources of Northeastern Mexico," U.S. Geological Survey Open-File Report 81-1143 (Washington, D.C.: Department of the Interior, 1981), p. 7.

5. *Ibid.,* pp. 3, 6.

6. Modified from Masters and Peterson, "Assessment of . . . Northeastern Mexico."

7. Francisco O. Viniegra and Carlos Castillo-Tejero, *Golden Lane Fields, Veracruz, Mexico: Geology of Giant Petroleum Fields* (Tulsa, Okla.: American Association of Petroleum Geologists, 1970), p. 311.

8. Robert T. Tippee, "Exploration, Development, Production Activity Mushrooming in Mexico," *Oil and Gas Journal* (August 24, 1981), p. 90.

9. "Giant Fields in Southeast Mexico," *Oil and Gas Journal* (July 20, 1981), p. 90.

10. Arthur A. Meyerhoff, "Geology of Reforma-Campeche Shelf," *Oil and Gas Journal* (April 21, 1980), pp. 121–123.

11. "Giant Fields in Southeast Mexico," pp. 89–92.

12. Shannon L. Matheny, Jr., "Big Secondary Recovery Waterflood Plans Are Taking Shape In Mexico," *Oil and Gas Journal* (September 29, 1980), p. 107.

13. "Giant Fields in Southeast Mexico."

14. *Ibid.*

15. Charles D. Masters and James A. Peterson, "Assessment of Conventionally Recoverable Petroleum Resources of Southeastern Mexico, Northern Guatemala, and Belize," U.S. Geological Survey Open-File Report 81-1144 (Washington, D.C.: Department of the Interior, 1981), p. 7.

16. *Ibid.*, pp. 3–6.

17. "Pemex Oil Flow Won't Top 2.9 Million B/D," *Oil and Gas Journal* (June 15, 1981), p. 55.

10
The Petroleum Prospects
of the Arabian-Iranian Basin

The great extracontinental closed downwarped sedimentary basin, of which the Persian Gulf is the center, contains broad, gentle geologic structures filled with the world's richest reserves of oil (see Figure 10.1). This vast basin is bounded on the north by the Taurus Mountains in Turkey, on the east by the mountains of Iran and Oman, and on the west by the uplands of Levant. The discovery and development of the enormous oil resources of the Arabian-Iranian basin have provided the supplies necessary to meet world (and especially European and Japanese) oil requirements during the post–World War II period. Sufficient resources remain for the region to continue as the world's major oil exporter for several decades into the future.

History

The founding of the oil industry in the Persian Gulf area dates back to 1908, when oil was discovered in southwest Iran. Iranian oil facilities and output were rapidly expanded during and after World War I, but production fell sharply in the early years of World War II. Recovery began in 1943 and 1944 with the reopening of supply routes to England, and oil was produced by what became the Anglo-Iranian Oil Company. In the postwar period, however, political difficulties arose between the company and the Iranian government. The conflict centered on government dissatisfaction with the financial terms of the petroleum concession and with the monopoly position of the company and its close association with the British government. The subsequent breakdown in relations between the state and the company led to the nationalization of the oil industry in 1951.[1]

320

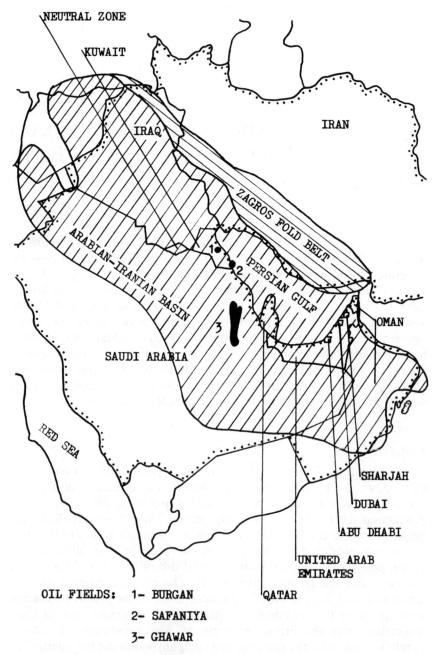

Figure 10.1. Map of Arabian-Iranian basin region.

Following nationalization, oil production dropped steeply. A compromise was not reached until 1954, when an agreement was signed between the Iranian government and a group of oil companies (including British Petroleum—formerly the Anglo-Iranian Oil Company) for exploitation and marketing rights to Iranian oil. The agreement recognized that the National Iranian Oil Company retained the sole ownership of all fixed assets of the Iranian oil industry. In 1973, Iran took charge of its oil industry and, under joint arrangements with overseas oil companies, generally continued the expansion of production that had resulted from the 1954 compromise. Continued expansion of Iranian oil production was made possible by a succession of discoveries of large petroleum accumulations. These discoveries included the supergiant Agha-Jari, Gach Saran, Marun, Ahwaz, and Rag-e Safid fields. Lately, the Iranian revolution and the war with Iraq have resulted in a precipitous decline in Iranian petroleum production.

Little oil development had been possible in the Gulf region during World War II, and at the war's conclusion only 7 fields were in production in the entire area, although 14 additional fields had been discovered by that time. The fields were located in Iran, Iraq, Qatar, Kuwait, and Saudi Arabia. Although little oil was being produced, it was by then evident that the Gulf would become a major petroleum-exporting region when adequate outlets became available. The postwar years' production expansion was coupled with a rapid increase in world oil demand.

With the nationalization of Iranian oil in the early 1950s and the resulting decline in production, Kuwait became the Gulf's leading oil producer. This was due to the dramatic expansion of output from the Burgan field (the world's second largest oil field originally containing 75 billion barrels of oil), which had been discovered in 1938 (see Figure 10.1). Kuwait held its position as the region's leading oil producer until 1965.[2]

Saudi Arabia rose to prominence as an oil-rich state later than most others in the Gulf. It has since, however, achieved preeminence as the largest single holder of oil in the world. In an atmosphere of competition between the established British (in Iran, Iraq, and Kuwait) and the incoming American oil companies, Saudi Arabia granted a concession to the Arabian American Oil Company (Aramco) in 1933. Early exploration produced the first discovery in 1935, but commercial oil accumulations were not located until 1938. The first exports were very modest and these modest levels of exports continued until after World War II. The discovery that transformed the prospects for the oil industry in Saudi Arabia was of Ghawar in 1948. This field proved to be the largest in the world (86 billion barrels), although it was originally thought to be several separate smaller fields. Another important discovery was of Safaniya, offshore in the Persian Gulf. It is the fourth-largest field in the world (31 billion barrels) and the world's largest offshore field. Other Saudi Arabian super-giants include: Berri, Manifa, Abu Sa'fah, Qatif, Shaybah, Abqaiq, Zuluf, Marjan, and Khurais.

A national oil company was established in Saudi Arabia in 1956 to supervise the exploration and production of petroleum resources outside of the Aramco concession. Since that time, oil production has become increasingly governed by the state-owned company, Petromin. In 1974, by agreement between the Saudi government and Aramco, Petromin purchased 60 percent participation in Aramco. It is anticipated that Aramco's assets in Saudi Arabia will eventually be completely nationalized. Saudi Arabia currently produces over 60 percent of Persian Gulf oil, far more than any other Gulf state, including Kuwait.

Geology

Underlying the Arabian-Iranian basin is a deep basement of Precambrian rock that outcrops on the Western Arabia plateau. The outcrops are Precambrian massifs, blocks of intensely deformed igneous and metamorphic rocks. On the south they form part of a crystalline mass extending into Yemen and South Yemen and on the west outcrop in portions of the Central Iranian plateau. The Precambrian rocks exhibit a complex history. They were deformed and stabilized and underwent several periods of erosion before the beginning of the Paleozoic era.[3] A shelf (or platform) area occurs to the north and northwest of the Arabia Precambrian massif, which is the result of intermittent, slow subsidence under shallow seas during the Paleozoic and Mesozoic, between Cambrian and Cretaceous times. A thick sequence of continental and shallow marine sediments forms this wide Arabian platform that dips gently east and northeast. The elimination of the ancient seaway began in the Late Cretaceous with orogenic movement that climaxed in late Tertiary time when the rocks in the deeper parts of the basin were folded to form the great mobile belt of the Zagros, Taurus, and Oman mountains.

The shelf can be divided into an almost flat homocline bordering the Arabia massif and an interior platform further to the east that forms a belt of more steeply dipping strata. The irregularities of dip in this area are related to the occurrence of a number of regional north-south anticlinal trends that are probably related to basement uplifts. Many major oil fields occur in structural closures on these trends. The most important of the trends is the En Nala or Hasa arch. Associated with this arch are many major oil fields, such as Ghawar, Abqaiq, Qatif, and Burgan, as well as Bahrein and Dukhan (Qatar) and some of the offshore Iranian fields.[4]

The mobile belt lies to the north and east of the shelf area. The sediments in the deeper parts of the basin were folded by tangential pressures during Late Cretaceous to late Tertiary time with the direction of movement northeast to southwest. The folds gradually broaden to the southwest. In northeast Iran a series of overthrusts occurs, and the mountain ranges are tightly folded and faulted. Thus, the oil fields of Iran and nearby Iraq tend to be elongated in a northeast-southwest orientation. The fields to the south, in Kuwait and Saudi Arabia, are broader and tend to be oriented north to south.

The oil fields vary in their structural and stratigraphic characteristics according to their position in the basin. The Jurassic reservoirs occur mainly in broad, gentle structures that may be associated with salt-related or basement-related uplifts at depth. These fields are located in Saudi Arabia and in other nearby states such as Qatar. Most of the oil production is from the Late Jurassic Arab formation composed of carbonates alternating with anhydrites. The sequence is capped with the thick Hith anhydrite. An ancient carbonate platform was attached to the western landmass. Fine lagoonal sediments were deposited in the west and north associated with anhydrite and calcareous sand (in the form of offshore bars). Deeper water mud accumulated in the east. Two general types of carbonate sediments were packstones (grain-supported, muddy carbonate clastics) and grainstones (mud-free carbonate clastics—necessarily grain-supported). The grainstones and regressive sandstones form the reservoir rocks, while some of the oil may have originated in the packstones. During a transgressive episode in the middle of the Jurassic, organically rich carbonates accumulated under anaerobic conditions in smaller basins on the platform. These deposits are the most likely source for most of the petroleum that has accumulated in the Jurassic reservoirs of Saudi Arabia.[5] Grainstones of equivalent age were deposited in adjoining areas of the platform. Jurassic sedimentation in the region concluded with four main depositional cycles. Each cycle consists of a shallowing marine carbonate sequence overlain by anhydrite and is composed of grainstones containing most of the oil accumulated in Jurassic reservoirs in Saudi Arabia. They are known as Arab A, B, C, and D reservoirs. The concluding evaporite unit, the Hith anhydrite, averages 500 feet in thickness and provides an extensive seal that has prevented the further upward migration of oil generated in Jurassic source rocks.[6]

The largest oil accumulations occur in the lowest grainstone cycle of the Arab formation (the Arab D reservoir). The largest single accumulation is in the Arab D reservoir in Saudi Arabia's Ghawar field. At Ghawar, the accumulation is approximately 140 miles long (in a north-south direction) and covers about 875 square miles. The oil column reaches a maximum of 1,300 feet. Most of the original porosity is retained in the reservoir.

During Early Cretaceous time, shallow-water marine carbonates with clastic textures were deposited on an extensive submarine platform. In the northeastern part of Saudi Arabia the basal Cretaceous unit locally contains zones that appear to be source rocks. During the middle Cretaceous period, cyclic sedimentation predominated with the deposition of nonmarine to marine clastics, which alternated with shallow-water marine carbonates. Regressive sandstones, which are prolific petroleum reservoirs in the super-giant Safaniya and in other Saudi Arabian fields, were deposited at this time. Oil also accumulated in the carbonate rocks that had been deposited. During the Late Cretaceous, predominantly shallow-water carbonates were deposited over the broad platform area. Tectonic movement on anticlinal structures occurred during this time, but additional structural growth also occurred during middle Tertiary time.[7]

The relatively large number of oil reservoirs in the region implies extensive vertical migration of petroleum from a more limited occurrence of rich source rocks. To reach the Jurassic Arab reservoirs, petroleum from the source rock must have moved upward through about 1,000 feet of intervening section. The presence of oil in the upper Arab reservoirs indicates that at least locally the Arab anhydrite separators are absent or that communication was achieved through fractures or fault offsets. The Hith anhydrite, however, appears to be an effective seal over the entire region with the exception of the northeast offshore area.

Geochemical evidence indicates that the petroleum in the Cretaceous offshore reservoirs in the northeast could have originated in Jurassic source rocks. To have migrated into these reservoirs, the petroleum would have had to move vertically as much as 5,000 feet. Upper Jurassic evaporites are poorly developed and discontinuous in this area, so it is possible that the extensive vertical movement of oil could have occurred. It is also possible that the trapped oil could have originated in Cretaceous source rocks in basinal areas in the northeast or east.[8]

In Saudi Arabia the extremely large oil accumulations in the Arabian shelf in Jurassic limestones can be attributed to: the widespread development of thermally mature organic-rich sediments underlying or adjacent to carbonate shelf sediments that have high porosities and permeabilities, a cycle of deposition concluding with a major regression during which the region was covered by continuous evaporite seals, and tectonic activity sufficient to create large structural traps but not intense enough to disrupt oil migration paths or evaporite caps. Also, the extraordinary horizontal extent of the structures and lithologic units in the basin permitted efficient horizontal migration. The traps with access to rich source rocks were filled because they drained such large source areas.

To the north, in Kuwait, the largest petroleum accumulation is Burgan, a gentle north-south elliptical dome with an area of about 135 square miles. Closure is at about 12,000 feet. Production comes from a number of highly permeable middle Cretaceous sandstones that are interbedded with shales. There is also a deeper Jurassic oil reservoir.

In Iraq, the super-giant Kirkuk field was discovered in 1925. It is a very large anticline whose principal reservoir is a reefal, highly permeable, late Eocene to Oligocene limestone. Oil is also found in younger limestones and older Cretaceous reservoirs. The overlying salt and anhydrite beds have formed an adequate but imperfect seal that has allowed oil and gas seeps near the crest of the southern part of the field. The petroleum is thought to have originated in the underlying Cretaceous dolomites and limestones.

The principal tectonic feature of Iran is the folded Zagros mountain belt that resulted from a Late Cretaceous–Tertiary orogeny in a mobile trough containing Triassic to Cenozoic sediments. Paralleling the mountain belt, a series of long, sinuous asymmetrical folds with a northwest-southeast orientation occurs that contrasts with the north-south trends of the principal Arabian shelf structures. The major reservoir in nearly all of Iranian oil fields is the Asmari formation, a reef-type limestone of Oligocene–early Miocene age. The shallow-water deposit is about 1,000 to 1,600 feet thick and has a characteristically high secondary permeability due to fracturing.[9] It is comformably overlain by evaporites that provide generally efficient seals, although petroleum seeps are common throughout the region. The Asmari limestone grades into continental sandstones to the west. These sandstone beds contain oil in some fields. Middle and Late Cretaceous limestones of southwest Iran also contain oil accumulations but are not as prolific as the Asmari limestone.

There are two structural trends offshore. The geologically younger trend parallels the northwest-southeast axes of the Iranian onshore Tertiary fields, and the older trend parallels the north-south axes of the Arabian Mesozoic fields. The reservoirs are the Oligocene–Miocene Asmari and the middle Cretaceous Sarvak limestones. Further south, production is from Lower Cretaceous and Jurassic Arab zone reservoirs.

There is an enormous gas potential in the Arabian-Iranian basin. The potential for gas associated with the major structures is huge.[10] The Permian Khuff formation underlies most of the Persian Gulf area and is an important gas-bearing horizon. It forms the reservoir in the super-giant North Dome gas field of offshore Qatar and in the onshore

TABLE 10.1   Total Original Recoverable Petroleum Resources in the
Arabian-Iranian Basin (including the Zagros Fold Belt)

	Oil *(billion barrels)*	*Natural Gas* *(trillion cubic feet)*
Cumulative production	124.814	data not available
Proved reserves	290.830	624.9
Inferred reserves	86.909	275.2
Undiscovered resources (mean)	174.5 (range 72–337)	849.5 (range 299–1,792)
Total	677.053	1,749.6+
1980 production	6.543	1.655
Reserves/Production ratio	44:1	378:1

Sources: Oil and Gas Journal (December 29, 1980, and December 28, 1981);
World Oil (August 1981, and February 1982); Charles D. Masters, "Assessment
of Conventionally Recoverable Petroleum Resources of the Persian Gulf Basin
and Zagros Fold Belt (Arabian-Iranian Basin)," U.S. Geological Survey Open-File
Report 81-986 (Washington, D.C.: Department of the Interior, 1981), p. 7; and
T. J. Stewart-Gordon, "Europe Sees Gas as Top Energy Source by 2000," *World
Oil* (April 1982), pp. 181–187.

Dunkhan field. It also produces nonassociated gas in Bahrain and in
the United Arab Emirates. Associated gas has been a problem in the
Gulf area, as much of the gas produced along with the oil has had to
be flared. Still, gas-gathering projects are well under way in Saudi Arabia,
Kuwait, Iraq, Bahrain, Oman, and the United Arab Emirates. Should
the demand for Gulf gas rise, either for export or internal use, there
is an enormous potential throughout the region either in the Permian
Khuff or in the Lower Cretaceous Thamama formations. The drilling
of these formations should result in a number of major discoveries of
nonassociated gas.

Oil and Natural Gas Production,
Reserves, and Resources

Table 10.1 shows production, reserves, and resources data for the
Arabian-Iranian basin.[11] (Individual countries are listed in Tables 11.1
and 11.4 in Chapter 11.) The reserve estimates for the Gulf region are
approximate due to the underdevelopment of many of the major fields.
There have been recent marked fluctuations in official reserve estimates
and considerable differences of opinion on recoverability. The inferred
reserves in the tables have been estimated by assuming that the higher

of the published reserve estimates include potential field growth. The region is so dominated by large structures that there is a lack of knowledge of smaller structure occurrence. Many wells that were tested at several thousand barrels of oil per day were plugged.[12]

The Arabian-Iranian basin contains by far the largest volume of oil of any region in the world. About 40 percent of this oil is in Saudi Arabia. A geological assessment of the undiscovered oil resources of the Gulf region indicates that about three-quarters of the recoverable oil has been found. The assessment presents a range of resource estimates, the most optimistic of which still indicates that over half of the region's total recoverable oil has already been discovered. The best potential for undiscovered oil is thought to be in Iraq, with good discovery prospects also remaining in Saudi Arabia and Iran. The smaller states, including Kuwait, have much more modest possibilities for additional discoveries.

Gas resources, on the other hand, have scarcely been tested in the Gulf region, and the gas potential is enormous. The greatest gas reserves and the greatest potential for undiscovered gas are in Iran. Saudi Arabia and the United Arab Emirates also have excellent gas prospects. Qatar has sizable gas reserves but poor prospects for additional gas discoveries.

Both oil and gas could be produced from the region at much higher rates, but there is little incentive to deplete resources faster and drive down prices. Sufficient oil resources remain for the Gulf region to continue as the major exporter of oil for many decades. Natural gas could have a similarly bright future as a commodity if an export market can be developed, either by means of a pipeline or as liquid natural gas.

Notes

1. Keith McLachlan, *Oil in the Persian Gulf Area: The Persian Gulf States* (Baltimore, Md.: Johns Hopkins University Press, 1980), p. 199.

2. *Ibid.*, p. 202.

3. Saleh M. Billo, "Future Petroleum Resources Seen Great in Saudi Arabia," *Oil and Gas Journal* (January 1, 1979), p. 99.

4. E. N. Tiratsoo, *Oilfields of the World,* 2nd ed. (Houston, Texas: Gulf Publishing Company, 1976), pp. 143–146.

5. M. G. Ayres, M. Bilal, R. W. Jones, L. W. Slentz, M. Tartir, and A. O. Wilson, "Hydrocarbon Habitat in Main Producing Areas: Saudi Arabia," *American Association of Petroleum Geologists Bulletin* (January 1982), pp. 1–9.

6. *Ibid.*, p. 3.

7. *Ibid.*

8. *Ibid.*, p. 8.

9. Tiratsoo, *Oilfields of the World,* p. 156.

10. Charles D. Masters, "Assessment of Conventionally Recoverable Petroleum Resources of the Persian Gulf Basin and Zagros Fold Belt (Arabian-Iranian Basin)," U.S. Geological Survey Open-File Report 81-986 (Washington, D.C.: Department of the Interior, 1981), p. 7.

11. Sources: *Oil and Gas Journal* (December 29, 1980, and December 28, 1981); *World Oil* (August 1981 and February 1982); Masters, "Assessment of . . . the Persian Gulf Basin and Zagros Fold Belt (Arabian-Iranian Basin)"; and T. J. Stewart-Gordon, "Europe Sees Gas As Top Energy Source By 2000," *World Oil* (April 1982), pp. 181–187.

12. Masters, "Assessment of . . . the Persian Gulf Basin and Zagros Fold Belt (Arabian-Iranian Basin)."

11
Worldwide Petroleum

Oil

The first 200 billion barrels of world oil was produced in 109 years, 1859 to 1968; the second 200 billion barrels was produced in 10 years, 1968 to 1978; and an additional 45 billion barrels was produced from 1978 to the end of 1980. Petroleum exploration has extended throughout the world, and current knowledge of world oil occurrence is extensive as well as intensive. Geologic understanding and exploration technology have progressed over the past several decades and are capable of distinguishing, in general, between geologically favorable and unfavorable conditions prior to drilling. Thus, only a relatively few exploratory holes may be necessary to indicate whether a region is likely to contain significant amounts of petroleum. As modern petroleum exploration is an efficient process, it is probable that most oil in a region will be found by the first 25 to 200 new field exploratory wells, if giant fields exist. This number of wells may be exceeded if there are very many major prospects or if exploration drilling patterns are controlled by political or technological constraints.[1]

Table 11.1 contains world oil production, reserves, and resources data, including super-giant fields, number of producing wells, undiscovered resources, and ultimate recovery (if known).[2] The table illustrates that the Middle East (with 26 super-giant fields) dominates world oil, with 52 percent of proved world reserves.

Although about 75 percent of the world's exploratory wells have been drilled in the United States, the rest of the world is not necessarily underexplored. As well over half of the oil in most sedimentary basins exists in a small number of large, relatively easy to find fields, the exhaustive exploration of a producing basin may not significantly increase total production from that basin, even though numerous small pools may be discovered.[3] The real contribution of such small fields may be

TABLE 11.1   World Oil Production, Reserves, and Resources for 1980
(in billion barrels)

Country Super-Giant Fields	Production Producing Wells (w)	Reserves Proved (p) Inferred (i)		Reserves/ Production Ratio	Cumulative Production (c) Undiscovered Resources (u) Ultimate Recovery (ur)	
North and Central America						
Canada	.538	5.986	(p)	11:1	10.288	(c)
	20,721 (w)	.414	(i)			
Mexico	.717	44.000	(p)	61:1	8.332	(c)
Cantarell	3,384 (w)	13.000	(i)		79.3	(u)
Bermudez					144.632	(ur)
United States	2.975	27.046	(p)	9:1	123.675	(c)
Prudhoe Bay	526,855 (w)	26.300	(i)		82.6	(u)
East Texas					259.621	(ur)
Barbados	.0004	.0008	(p)	2:1	.002	(c)
	34 (w)					
Guatemala	.0011	.020	(p)	18:1	.004	(c)
	3 (w)					
Trinidad &	.079	.650	(p)	8:1	2.032	(c)
Tobago	3,349 (w)	.050	(i)		2.1	(u)
					4.832	(ur)
South America						
Argentina	.179	2.457	(p)	14:1	3.567	(c)
	6,822 (w)					
Bolivia	.011	.112	(p)	10:1	.245	(c)
	292 (w)	.056	(i)			
Brazil	.070	1.300	(p)	19:1	1.198	(c)
	1,615 (w)	.018	(i)			
Chile	.011	.400	(p)	36:1	.269	(c)
	380 (w)	.180	(i)			
Colombia	.046	.516	(p)	11:1	2.188	(c)
	2,123 (w)	.284	(i)			
Ecuador	.084	1.050	(p)	12:1	.728	(c)
	477 (w)	.050	(i)			
Peru	.070	.628	(p)	9:1	1.043	(c)
	2,782 (w)	.022	(i)			
Venezuela	.787	17.950	(p)	23:1	36.441	(c)
Bolivar Coastal	5,042 (w)	1.669	(i)		23.0	(u)
					79.060	(ur)
Total Western	5.5685	102.1158	(p)	18:1	190.012	(c)
Hemisphere	573,879 (w)	42.043	(i)			
Western Europe						
Austria	.011	.123	(p)	11:1	.606	(c)
	1,120 (w)	.007	(i)			

TABLE 11.1 *(continued)*

Country Super-Giant Fields	Production Producing Wells (w)	Reserves Proved Inferred	(p) (i)	Reserves/ Production Ratio	Cumulative Production (c) Undiscovered Resources (u) Ultimate Recovery	(ur)
Denmark	.002	.450	(p)	225:1	.019	(c)
	11 (w)	.025	(i)			
France	.009	.040	(p)	4:1	.381	(c)
	296 (w)	.063	(i)			
West Germany	.033	.431	(p)	13:1	1.344	(c)
	2,957 (w)	.019	(i)			
Greece		.055	(p)			
		.095	(i)			
Italy	.014	.362	(p)	26:1	.260	(c)
	116 (w)	.278	(i)			
Netherlands	.009	.123	(p)	14:1	.351	(c)
	480 (w)	.162	(i)			
Norway	.204	5.500	(p)	27:1	.881	(c)
	100 (w)	1.120	(i)			
Spain	.009	.222	(p)	25:1	.088	(c)
	31 (w)	.128	(i)			
Sweden	.044					
United Kingdom	.577	8.209	(p)	14:1	2.154	(c)
	290 (w)	6.591	(i)			
Yugoslavia	.029	.265	(p)	9:1		
		.021	(i)			
North Sea (south of 62° north latitude)					13.0	(u)
Total Western Europe	.941 5,401 (w)	15.780 8.509	(p) (i)	17:1	6.084	(c)

Eastern Europe

Country Super-Giant Fields	Production Producing Wells (w)	Reserves Proved Inferred	(p) (i)	Reserves/ Production Ratio	Cumulative Production (c) Undiscovered Resources (u) Ultimate Recovery	(ur)
Albania	.023	.154		7:1		
Bulgaria	.002	.029		16:1		
Czechoslovakia	.001	.013		13:1		
East Germany	.0004	.030		75:1		
Hungary	.015	.188		13:1		
Poland	.003	.019		6:1		
Rumania	.085	.850		10:1		
Soviet Union	4.390	63.000	(p)	14:1	63.0+	(c)
Samotlor	50,000 (w; est.)	2.000	(i)		115.0+	(u)
Romashkino					243.0+	(ur)
Total Eastern Europe	4.5194 50,000+(w)	64.283	(p)	14:1	63.0+	(c)

TABLE 11.1 *(continued)*

Country Super-Giant Fields	Production Producing Wells (w)	Reserves Proved Inferred	(p) (i)	Reserves/ Production Ratio	Cumulative Production (c) Undiscovered Resources (u) Ultimate Recovery (ur)	
Middle East						
Bahrein	.018	.214	(p)	12:1	.7	(c)
	242 (w)	.011	(i)		0	(u)
					.925	(ur)
Iran	.468	39.935	(p)	85:1	29.9	(c)
Ahwaz	547 (w)	17.565	(i)		26.0	(u)
Marun					113.4	(ur)
Gach Saran						
Agha-Jari						
Rag-e Safid						
Iraq	.920	30.000	(p)	33:1	15.704	(c)
Rumaila	250 (w)	4.100	(i)		78.0	(u)
Kirkuk					127.804	(ur)
Zubair						
Israel	.0002	.001	(p)	5:1	.016	(c)
	9 (w)					
Kuwait	.613	67.930	(p)	111:1	22.143	(c)
Burgan	796 (w)	3.333	(i)		4.0	(u)
Randhatain-Sabriya					97.406	(ur)
(includes one-half neutral zone data)						
Oman	.103	2.340	(p)	23:1	1.543	(c)
	269 (w)	.300	(i)		2.0	(u)
					6.183	(ur)
Qatar	.173	3.587	(p)	21:1	3.227	(c)
	124 (w)				.5	(u)
					7.314	(ur)
Saudi Arabia	3.622	116.414	(p)	32:1	44.139	(c)
Ghawar	939 (w)	51.616	(i)		57.0	(u)
Safaniya-Khafji					269.169	(ur)
Berri						
Abqaiq						
Manifa						
Zuluf						
Fereidoon-Marjan						
Qatif						
Khurais						
Shaybah						
Abu Sa'fah						
(includes one-half neutral zone data)						
Syria	.060	1.605	(p)	27:1	.618	(c)
	179 (w)	.335	(i)			
Turkey	.019	.110	(p)	6:1	.372	(c)
	414 (w)	.165	(i)			

TABLE 11.1 *(continued)*

Country Super-Giant Fields	Production Producing Wells (w)	Reserves Proved Inferred	(p) (i)	Reserves/ Production Ratio	Cumulative Production (c) Undiscovered Resources (u) Ultimate Recovery	(c) (u) (ur)
United Arab	.626	30.410	(p)	49:1	7.458	(c)
Emirates	424 (w)	6.397	(i)		7.0	(u)
Abu Dhabi					51.265	(ur)
Zakum						
Bu Hasa						
Asab						
Umm Sharf						
Bab						
Dubai						
Sharjah						
Total Middle	6.6222	292.546	(p)	44:1	125.820	(c)
East	4,193 (w)	83.822	(i)			
Africa						
Algeria	.366	8.200	(p)	22:1	5.802	(c)
Hassi Messaoud	1,000 (w)	3.575	(i)			
Angola	.051	1.200	(p)	24:1	.635	(c)
	188 (w)	.175	(i)			
Cameroun	.021	.200	(p)	10:1	.054	(c)
	49 (w)	.070	(i)			
Congo Republic	.022	.660	(p)	30:1	.154	(c)
	119 (w)	.045	(i)			
Egypt	.213	2.900	(p)	14:1	2.426	(c)
	484 (w)	.300	(i)			
Gabon	.065	.450	(p)	7:1	.879	(c)
	242 (w)	.021	(i)			
Ghana	.001	.006	(p)	6:1	.002	(c)
Ivory Coast	.001	.050	(p)	50:1	.002	(c)
Libya	.659	23.000	(p)	35:1	13.669	(c)
Sarir	886 (w)	2.788	(i)			
Amal						
Nigeria	.758	11.295	(p)	15:1	8.708	(c)
	1,451 (w)	5.405	(i)		8.4	(u)
					33.808	(ur)
Tunisia	.043	1.652	(p)	38:1	.469	(c)
	87 (w)	.529	(i)			
Zaire	.007	.115	(p)	16:1	.041	(c)
	13 (w)	.015	(i)			
Total Africa	2.207	49.728	(p)	23:1	32.841	(c)
	4,519 (w)	12.923	(i)			

TABLE 11.1 *(continued)*

Country Super-Giant Fields	Production Producing Wells (w)	Reserves Proved (p) Inferred (i)		Reserves/ Production Ratio	Cumulative Production (c) Undiscovered Resources (u) Ultimate Recovery (ur)	
Asia-Pacific						
Australia	.147	2.360	(p)	16:1	1.582	(c)
	403 (w)	.140	(i)			
Brunei	.080	1.294	(p)	16:1	1.645	(c)
	578 (w)	.416	(i)			
Burma	.011	.030	(p)	3:1	.422	(c)
	445 (w)	.006	(i)			
China Ta-Ching	.773	19.000	(p)	25:1		
		1.500	(i)			
India	.070	2.409	(p)	34:1	.950	(c)
	1,600 (w)	.171	(i)			
Indonesia	.576	9.800	(p)	17:1	9.200	(c)
	3,700 (w)				16.0	(u)
					35.0	(ur)
Japan	.003	.031	(p)	10:1	.109	(c)
	388 (w)	.021	(i)			
Malaysia	.102	1.800	(p)	18:1	.710	(c)
	230 (w)	1.200	(i)			
New Zealand	.003	.083	(p)	28:1	.028	(c)
	16 (w)					
Pakistan	.004	.197	(p)	49:1	.092	(c)
	19 (w)	.275	(i)			
Philippines	.002	.020	(p)	10:1	.013	(c)
	4 (w)	.171	(i)			
Taiwan	.001	.012	(p)	12:1	.016	(c)
	82 (w)					
Total Asia-Pacific	1.772	37.036	(p)	21:1	14.767	(c)
	7,465 (w)	3.900	(i)			
Total World	21.6301	561.4888	(p)	26:1	432.524+	(c)
	645,457+(w)	151.197	(i)			

Note: Undiscovered oil resources in regions that have been assessed total 513.9+ billion barrels. Inferred reserves are estimated by assuming that the high side of published reserve estimates includes potential field growth.

Source: Oil and Gas Journal (December 29, 1980, and December 28, 1981); *World Oil* (August 1981 and February 1982); *Petroleum Economist* (August 1981); T. J. Stewart-Gordon, "Europe Sees Gas as Top Energy Source by 2000," *World Oil* (April 1982), pp. 181–187; and "USGS Issues Reports on World Oil, Gas Potential," *Oil and Gas Journal* (December 7, 1981), pp. 264–268.

in the future, at the end of the exploration process, during the period of transition to other energy forms.

Worldwide, some 600 sedimentary basins are known to occur. About 200 of these have had little or no exploration for petroleum. They are not completely unknown geologically, as sufficient work has been done to indicate their dimensions, type of sediment, and gross structural aspects. Almost all of these frontier basins are in polar regions, under relatively deep waters offshore, or in the remote interiors of continents. Exploration in a few others has been restricted for political reasons. Another 240 basins have been explored to some extent without yielding commercial discoveries. The remaining 160 basins are commercially productive in petroleum.[4] Of these, 7 contained more than 24 billion barrels of known recoverable oil, 8 contained between 12 and 24 billion barrels, the next 11 largest contained between 6 and 12 billion barrels, 24 contained between 2 and 6 billion barrels, and the remaining 110 contained a total of 52 billion barrels.[5] This information is used in Table 11.2 to estimate the total recoverable oil in the 200 frontier basins at 294 billion barrels. Total world recoverable oil is estimated to be 1,953.1 billion barrels, as shown in Table 11.3.

The estimated undiscovered recoverable oil resources in the under-explored frontier basins are likely to be too high for several reasons. Even with the exclusion of the Arabian-Iranian basin (as it is very unlikely that another such basin exists), the quality of the frontier basins is not expected to equal that of the explored basins. Many of the frontier offshore basins are pull-apart basins and the remote continental basins are likely to be interior basins, neither of which has been found to be very productive. In addition, the costs involved in the exploration of these frontier basins will not allow as intensive development as occurs in more favorable environments. Thus, smaller fields will not be as economically viable and overall recovery efficiency will not be as high. On the other hand, as not all productive areas have been assessed, the assessed undiscovered resources estimate is too low. Given a high and low bias to the resource estimates, a total world recoverable oil stock of around 2,000 billion barrels is a reasonable estimate, given the state of current knowledge. Also, well over half of the world's potential resources are likely to be in currently productive basins, as indicated in Table 11.3. Table 11.3 also indicates that 59 percent of the world's oil has already been discovered and that 22 percent of world oil stocks have been consumed. Current consumption rates are about 1 percent of the total per year.

Heavy oil may be an additional source of energy, although it is currently being produced and often counted in oil reserve estimates. It appears that the amount of in-place heavy crude oil in the world

TABLE 11.2   Estimated Undiscovered Recoverable Oil in 200 Underexplored Frontier Basins (in billion barrels)

Explored basins	400	basins
Explored basins with commercial production	160	(40 percent)
40 percent of frontier basins (200)	80	(may be productive)
4 percent of productive basins average 36 billion barrels of known recoverable oil (excluding Arabian-Iranian basin) 4 percent of frontier basins (3 basins x 36)	108	billion barrels
5 percent of productive basins average 17 billion barrels of known recoverable oil 5 percent of frontier basins (4 basins x 17)	68	billion barrels
7 percent of productive basins average 9 billion barrels of known recoverable oil 7 percent of frontier basins (6 basins x 9)	54	billion barrels
15 percent of productive basins average 3 billion barrels of known recoverable oil 15 percent of frontier basins (12 basins x 3)	36	billion barrels
The remaining 69 percent of productive basins average 0.5 billion barrels of known recoverable oil The 55 remaining frontier basins x 0.5	28	billion barrels
Total estimated oil in the 200 frontier basins based on the experience of the 160 productive basins and the 240 unproductive basins	294	billion barrels

TABLE 11.3   Estimated Total World Recoverable Oil (in billion barrels)

Cumulative production	432.5+
Proved reserves	561.5
Inferred reserves	151.2
Assessed undiscovered resources	513.9+
Estimated undiscovered resources	294.0−
Total world recoverable oil	1,953.1

may be about equal to the amount of lighter crude; distribution is such that the heavy crudes occur mostly in the Western Hemisphere (especially in Venezuela and Canada) and the lighter crudes occur mostly in the Eastern Hemisphere. The heavier crudes are a much less desirable energy resource than the lighter crudes, as they are much more costly to extract and to process. The total original heavy oil in place in the United States has been estimated at 125 billion barrels,[6] but the total oil estimated to be recoverable from these known heavy oil reservoirs (using thermal enhanced recovery methods) is only 7.5 billion barrels.[7] Heavy oil fields are produced at slower rates than conventional oil fields, and on the average about 1 barrel of oil is combusted to produce the steam necessary to net 2 barrels of heavy oil. This reduces the recoverable oil in a heavy oil reservoir by a factor of one-third compared to a conventional oil reservoir.[8] Low recovery efficiencies and expensive production methods will inhibit heavy oil utilization and prevent it from providing a volume of liquid petroleum similar to that of the lighter oils.

Natural Gas and Natural Gas Liquids

As recently as 1960, natural gas was a nuisance by-product of oil production in many areas of the world, something to be separated from the crude stream and gotten rid of as cheaply as possible (often by flaring). After twenty years of rapidly increasing proved gas reserves throughout the world (a period that included two crude oil supply crises), natural gas has become an important energy source. Table 11.4 contains world natural gas production, reserves, and resources information. It can be seen from the table that the Soviet Union is the world's leading region for gas resources and reserves.

World-class giant gas fields (larger than 3 trillion cubic feet) represent less than 1 percent of the world's gas fields but contain (along with associated gas in giant oil fields) about 80 percent of the world's proved and produced gas reserves. Oil is derived mainly from marine or lacustrian source rocks, and gas is derived from both land plant and marine organic matter. Thus, all source rocks have the potential for gas generation. Many of the source rocks for large gas deposits appear to be associated with the upper Paleozoic worldwide coal occurrence.

Large gas fields have been found above the oil window as a result of a biogenic origin. Thermal gas occurs below the oil window, and gas thus becomes the main target of deep drilling. Therefore, in most basins the vertical potential (and basin sediment volume) available for gas generation and accumulation exceeds that of oil. About a quarter of the giant gas fields are related to a shallow biogenic origin, but most

TABLE 11.4   World Natural Gas Production, Reserves and Resources for 1980
(in trillion cubic feet)

Country	Production	Reserves Proved Inferred	(p) (i)	Reserves/ Production Ratio	Undiscovered Resources
North and Central America					
Canada	2.46	87.3	(p)	35:1	
		.8	(i)		
Mexico	1.06	75.4	(p)	71:1	180.5
United States	19.33	188.16	(p)	10:1	593.8
		174.4	(i)		
Trinidad & Tobago	.085	7.06	(p)	83:1	20.0
		4.94	(i)		
South America					
Argentina	.279	20.3	(p)	73:1	
		1.7	(i)		
Bolivia	.082	4.2	(p)	51:1	
Brazil	.030	1.5	(p)	50:1	
Chile	.100	2.5	(p)	25:1	
Colombia	.098	4.6	(p)	47:1	
		1.4	(i)		
Ecuador	.002	4.0	(p)	2000:1	
Peru	.040	1.1	(p)	28:1	
		.2	(i)		
Venezuela	.528	42.0	(p)	80:1	59.0
		2.1	(i)		
Total Western Hemisphere	24.094	438.12 185.54	(p) (i)	18:1	
Western Europe					
Austria	.067	.424	(p)	6:1	
Denmark		2.82	(p)		
France	.266	3.3	(p)	12:1	
		2.7	(i)		
West Germany	.669	6.0	(p)	9:1	
		.5	(i)		
Greece		4.0	(p)		
Ireland	.030	.95	(p)	32:1	
Italy	.421	3.5	(p)	8:1	
		2.7	(i)		
Netherlands	3.087	57.4	(p)	19:1	
		4.6	(i)		
Norway	.886	29.3	(p)	33:1	
		13.4	(i)		
Spain		3.5	(p)		
United Kingdom	1.292	24.9	(p)	19:1	
Yugoslavia	.064	1.5	(p)	23:1	
		.4	(i)		

TABLE 11.4   *(continued)*

Country	Production	Reserves Proved Inferred	(p) (i)	Reserves/ Production Ratio	Undiscovered Resources
North Sea (south of 62° north latitude)					102
Total Western Europe	6.782	136.644 (p) 24.3 (i)		20:1	
Eastern Europe					
Albania		.35	(p)		
Bulgaria	.005	.18	(p)	36:1	
Czechoslovakia	.021	.46	(p)	22:1	
East Germany	.303	2.82	(p)	9:1	
Hungary	.216	4.06	(p)	19:1	
Poland	.224	4.41	(p)	20:1	
Rumania	.995	4.77	(p)	5:1	
Soviet Union	15.23	1,046.5 (p) 113.5 (i)		69:1	1,090+
Total Eastern Europe	16.994	1,063.55 (p) 113.5 (i)		63:1	
Middle East					
Bahrain	.097	9.0	(p)	93:1	0
Iran	.292	377.7 (p) 107.3 (i)		1293:1	560
Iraq	.062	27.5	(p)	444:1	31
Israel	.008	.1	(p)	12:1	
Kuwait	.240	33.3	(p)	139·1	11
Oman	.021	2.5 (p) 1.7 (i)		119:1	8
Qatar	.184	60.0 (p) 86.0 (i)		326:1	.5
Saudi Arabia	.530	92.8 (p) 19.6 (i)		175:1	201
Syria	.007	1.5 (p) 1.7 (i)		214:1	
Turkey		.5	(p)		
United Arab Emirates Abu Dhabi Dubai Sharjah	.229	22.1 (p) 60.6 (i)		97:1	38
Total Middle East	1.670	627.0 (p) 276.9 (i)		375:1	849.5

TABLE 11.4 *(continued)*

Country	Production	Reserves Proved Inferred	(p) (i)	Reserves/ Production Ratio	Undiscovered Resources
Africa					
Algeria	.380	132.0	(p)	347:1	
Angola	.009	1.0	(p)	111:1	
Congo Republic		2.0	(p)		
Egypt	.077	3.0	(p)	39:1	
Gabon	.003	.5	(p)	167:1	
Libya	.183	24.0	(p)	131:1	
Morocco	.003	.04	(p)	13:1	
Nigeria	.038	41.4	(p)	1089:1	65.0
Tunisia	.013	4.3	(p)	331:1	
		1.3	(i)		
Zaire		.05	(p)		
Total Africa	.706	208.29	(p)	295:1	
		1.3	(i)		
Asia-Pacific					
Afghanistan	.145	2.2	(p)	15:1	
Australia	.315	30.0	(p)	95:1	
Bangladesh	.049	8.0	(p)	163:1	
		1.0	(i)		
Brunei	.316	7.4	(p)	23:1	
Burma	.012	.18	(p)	15:1	
China	.734	24.5	(p)	33:1	
		.5	(i)		
India	.057	9.3	(p)	163:1	
		2.7	(i)		
Indonesia	1.2	27.4	(p)	23:1	42.0
Japan	.078	.67	(p)	9:1	
Malaysia	.036	15.0	(p)	417:1	
		9.1	(i)		
New Zealand	.044	6.1	(p)	139:1	
Pakistan	.286	15.1	(p)	53:1	
		.7	(i)		
Taiwan	.066	.83	(p)	13:1	
Thailand		8.0	(p)		
Total Asia-Pacific	3.338	154.68	(p)	46:1	
		14.0	(i)		
Total World	53.584	2,628.284	(p)	49:1	
		615.54	(i)		

Note: Undiscovered gas resources are indicated that have been assessed, a total of 3,001.8+ trillion cubic feet (expected value). Inferred reserves are estimated by assuming that the high side of published reserve estimates includes potential field growth.

Source: See Table 11.1.

of the giant deposits are located at intermediate or deeper zones where higher temperatures and older reservoirs (often carbonates) exist.[9]

Similarly sized oil traps contain more recoverable energy (on a Btu basis) than gas traps, even though conventional recovery of gas is much more efficient than conventional oil recovery. This is due to differences in the physical properties of gas and oil. Gas displays an initial low concentration and high dispersibility, making adequate cap rocks very important. It appears that, in general, relatively more gas per unit volume of sediments occurs in the larger basins. Intracontinental composite and complex basins, downwarp troughs, and delta basins appear to be gas prone. Pull-apart basins have also exhibited gas tendencies. It is possible that larger sediment volume and extensive carbonate sequences and evaporite seals, often typical of some of these basins, may result in the migration and entrapment of greater volumes of gas than in the smaller basins.

When the generation and migration of gas is considered, the extensive vertical gas zone includes shallow biogenic gas, intermediate dissolved gas of the oil window, and deep thermal gas. This large vertical habitat of gas and the additional availability of source material, when added to the past prospecting emphasis for oil, indicate that the world's ultimate resources of conventionally recoverable gas will approach those of oil.[10] On a Btu basis, this would amount to 11,328 trillion cubic feet. This is a large amount (some indications suggest one-third gas to two-thirds oil), but it can be seen from Table 11.4 that 6,246 trillion cubic feet of gas have already been accounted for as reserves or resources. Of that amount, some 3,244 trillion cubic feet have actually been discovered. What is not known is cumulative production, especially because significant amounts of gas have been flared in the past. Data from the United States, Indonesia, and Mexico indicate that more than 613 trillion cubic feet have been produced. Added to the discovered amount, about 3,857 trillion cubic feet of gas is accounted for, or about 34 percent of the total. If the assessed undiscovered gas is added, 61 percent of the estimated world total recoverable conventional natural gas is accounted for. The additional gas is likely to be found below Arctic permafrost, in bypassed gas discoveries, deep below the oil window, and, of course, in new and nonassessed basins.

An unknown amount of this total gas has been produced or has been (and is being) flared or otherwise wasted. (Approximately 10 percent of world gas production was lost at the wellhead in 1980 by flaring. The three years previous to 1980 averaged about 11 percent of gas production flared.[11] Historically, Saudi Arabia, Nigeria, Iran, Iraq, and Algeria have flared the most gas. The flaring of associated gas in these countries is a by-product of oil production. Much of the gas

TABLE 11.5   Estimated Total World Ultimate Recovery of Conventional
Petroleum

Oil	1,953	billion barrels
Natural gas liquids	428	billion barrels
Natural gas	11,328	trillion cubic feet

produced is reinjected, but what cannot be reinjected is flared. The remote location of many of these wells makes the recovery of gas expensive, but rising petroleum costs will provide an incentive to further reduce flaring in the future.

A prudent projection of gas production from unconventional geologic sources is that it will supplement rather than replace conventional gas production. If the more available unconventional gas sources (tight gas reservoirs and black Devonian-type shales) are considered, there is an important but relatively modest recoverable unconventional gas resource. Recoveries of very large volumes of unconventional gas would depend upon major production from geopressured brines and gray Devonian-type shales, difficult-to-produce resources due to very diffuse gas content. All types of unconventional gas will be much less efficient to recover than conventional gas and will have to be produced at slower recovery rates.[12]

To determine the ultimate amount of recoverable gas liquids in the world, the average production of gas liquids in the United States (37.8 barrels per million cubic feet of produced gas in 1980) can be assumed for the world ultimate gas recovery of 11,328 trillion cubic feet. On that basis the ultimate recoverable gas liquids estimate in the world is 428 billion barrels. When added to the estimated world ultimate oil recovery of 1,953 billion barrels, the estimate for the ultimate world production of petroleum liquids is 2,381 billion barrels. These amounts are shown in Table 11.5.

Notes

1. Richard Nehring, "The Outlook For World Oil Resources," *Oil and Gas Journal* (October 27, 1980), pp. 170–175.

2. Tables 11.1 and 11.4 were compiled from information in *Oil and Gas Journal* (December 29, 1980, and December 28, 1981); *World Oil* (August 1981 and February 1982); *Petroleum Economist* (August 1981); T. J. Stewart-Gordon, "Europe Sees Gas as Top Energy Source by 2000," *World Oil* (April 1982), pp. 181–187; and "USGS Issues Reports on World Oil, Gas Potential," *Oil and Gas Journal* (December 7, 1981), pp. 264–268.

3. Joseph P. Riva, Jr., *Conventional Crude Oil Availability; U.S. Refineries:*

A Background Study (Report prepared for the Subcommittee on Energy and Power, House Committee on Interstate and Foreign Commerce [Commitee Print 96-IFC 54], U.S. Congress, Washington, D.C., July 1980), pp. 105–109.

4. Michel T. Halbouty and John D. Moody, "World Ultimate Reserves of Crude Oil" (Paper presented at the Tenth World Petroleum Congress, Bucharest, Romania, September 9–14, 1979).

5. Nehring, "The Outlook for World Oil Resources."

6. National Petroleum Council, *Enhanced Oil Recovery* (Washington, D.C., 1976), pp. 182–183.

7. *Ibid.*

8. Joseph P. Riva, "Production, Reserves, and Processing of Domestic and Foreign Heavy Crude Oils," in *Studies in Taxation, Public Finance and Related Subjects—A Compendium,* vol. 4 (Washington, D.C.: Fund for Public Policy Research, 1980), pp. 477–493.

9. H. D. Klemme, personal communication.

10. *Ibid.*

11. Robert B. Kalisch, "Flaring of Natural Gas," *Gas Energy Review* (May 1982), pp. 10–13.

12. Joseph P. Riva, Jr., "Unconventional Gas in the United States: Reserves, Resources, and Production Projections" (*Hearings* on natural gas supply outlook, Subcommittee on Fossil and Synthetic Fuels of the House Committee on Energy and Commerce [Serial No. 97-90], U.S. Congress, Washington, D.C., June 1 and 9, 1981), pp. 455–482.

Index